SINISA STANKOVIC
DR NEIL CAMPBELL
DR ALAN HARRIES

URBAN WIND ENERGY

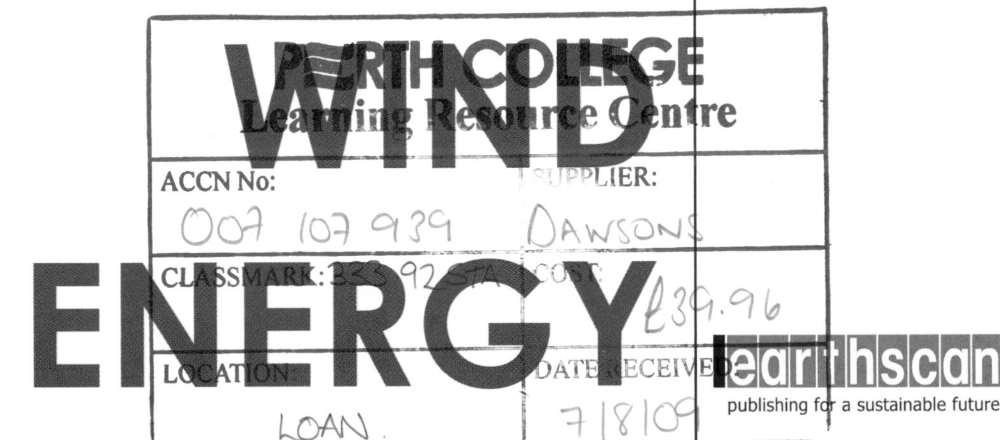

earthscan
publishing for a sustainable future

First published by Earthscan in the UK and USA in 2009

ISBN 978-1-84407-282-8

Typeset by Giles Bruce, Alan Harries and Katalin Pazstor
Cover design by Giles Bruce, image courtesy of Ecotricity

For a full list of publications please contact:

Earthscan
Dunstan House
14a St Cross St
London EC1N 8XA, UK
Tel: +44 (0)20 7841 1930
Fax: +44 (0)20 7242 1474
Email: earthinfo@earthscan.co.uk
Web: **www.earthscan.co.uk**

22883 Quicksilver Drive, Sterling, VA 20166-2012, USA

Earthscan publishes in association with the International Institute for Environment and Development

At Earthscan we strive to minimize our environmental impacts and carbon footprint through reducing waste, recycling and offsetting our CO_2 emissions, including those created through publication of this book. For more details of our environmental policy, see www.earthscan.co.uk.

This book was printed in the UK by Scotprint, an ISO 14001 accredited company. The paper used is FSC certified and the inks are vegetable based.

A catalogue record for this book is available from the British Library

Library of Congress Cataloging-in-Publication Data

Stankovic, Sinisa.
Urban wind energy / Sinisa Stankovic, Neil Campbell, and Alan Harries. -- 1st ed.
p. cm.
Includes bibliographical references and index.
ISBN 978-1-84407-282-8 (hardback)
1. Wind power. 2. City planning--Environmental aspects. I. Campbell, Neil. II. Harries, Alan. III. Title.
TJ820.S83 2009
333.9'2--dc22

2008052630

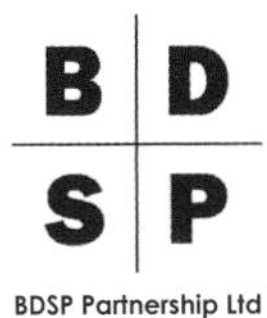

SINISA STANKOVIC
MSc Energy and Building Design, CEng, Fellow of CIBSE,
Director of BDSP Partnership

DR NEIL CAMPBELL
PhD Building Physics,
Associate of BDSP Partnership

DR ALAN HARRIES
PhD Fluid Dynamics,
Principal Wind Energy Consultant for BDSP Partnership

BDSP Partnership are international consulting engineers specializing in building services, sustainability, energy efficiency and masterplanning. From their inception in 1995, BDSP's core design principles have included an emphasis on reducing energy demand in buildings combined with high-efficiency energy supply and integration of renewable technologies.

Between 1995 and 1997, BDSP Partnership helped conceive one of the first concept buildings with integrated wind turbines as part of Project ZED (Towards Zero Emission Urban Development), which was part-funded by the European Commission (EC). In 1998, BDSP Partnership went on to establish EC Project WEB (Wind Energy in the Built Environment). This was one of the first research projects to begin to address the potential of urban wind energy. The project focused on the demonstration and evaluation of wind enhancement and integration techniques as well as assessing economic aspects and environmental impacts. Since the completion of this two-year research project, BDSP Partnership has been involved in a number of wind energy projects during a period which has seen a marked increase in worldwide interest in urban wind energy.

Contacting the Authors

In order to help evolve this text through subsequent editions, please contact the authors at **uwe@bdsp.com** to provide and discuss comments, feedback, omissions, corrections, case studies, technological advancement, lessons learned and proposals or visions.

Introduction

Towns and cities are essential to human development, with the majority of the population now living in urban areas. Buildings and their occupants are large consumers of finite energy and material resources and, therefore, major cumulative contributors to global environmental issues.

Cheap, secure, reliable energy supplies are key to the growth of modern industrial economies and there is often little incentive for large developers to explore the potential for local or decentralized electricity supplies. This is beginning to change in response to volatility in the fossil fuel markets, concerns about energy security, national and international commitments to sustainable growth and reduction of greenhouse gas emissions.

Many countries and regions have climates that are well suited to harnessing their wind resources for electricity generation. While there are now many projects where photovoltaic cells have been incorporated into buildings to take advantage of "free" solar energy, the number adopting wind energy is much smaller.

Integration of wind turbines into the built environment does pose challenges to be overcome – reductions in energy yields due to lower mean wind speeds in urban areas and associated environmental impacts due to their close proximity to people and property. However, in certain urban areas possessing suitable conditions there is potential for successful wind energy generation and small wind is expanding fast, particularly in the USA and UK.

This book helps identify and assess the potential of new wind energy projects in urban areas. These include: the potential owners (investors, developers, businesses, communities and homeowners), suppliers (manufacturers), distributors (utility companies and district network operators – DNOs), legal representatives (planners, policy-makers, funders and grant distributors), and designers/installers (architects, engineers and technicians). The following Parts aim to cover all relevant areas with the exception of: self-build turbines, detailed mechanical and electric equipment design, in-depth structural aspects, highly theoretical aspects and offshore wind.

1

2

3

4

5

PART 1 – Wind Energy in Context

The books begins by tackling the fundamental question: 'why should we be interested in generating energy from wind?' – and more specifically 'why in an urban context?'. The main drivers are identified and a series of wider contextual questions addressed to help establish the relative importance of urban wind energy in our energy futures.

PART 2 – Urban Wind Energy Potential

In this section, three main categories for wind energy integration in the built environment are considered: small wind and retrofitting, large-scale stand-alone turbines and building-integrated turbines (where the buildings are designed with wind energy in mind). A variety of case studies are presented and reviewed.

PART 3 – Urban Wind Energy Feasibility Study

How should one determine the viability of a potential project? This section examines the areas that make up a formal feasibility study - necessary when progressing an urban wind energy project through the planning permission application stage. This includes methods for wind resource estimation, predicting performance, evaluating environmental impacts and making economic assessments.

PART 4 – Turbine Technology

This section contains a detailed overview of wind turbine technology – including the fundamentals of horizontal and vertical axis turbine design and general wind energy yield enhancement techniques. A focus on practical implementation is maintained in relation to the pros and cons of working with the current available turbine technology but combined with a solid theoretical grounding.

PART 5 – Building-Integrated Wind Turbines

In the final part of this book, design issues related specifically to building-integrated wind energy are considered. This includes exploring generic integration techniques and the application of 'state-of-the-art' computational simulation to ensure energy production levels are maximized while keeping environmental impacts low.

Table of Contents

PART 4

Wind Energy in Context

Large-scale turbines in Taiwan
(Te-Wei Liu)

1

INTRODUCTION

ENERGY ENERGY ENERGY...

Wind energy generation is growing rapidly worldwide and will continue to do so for the foreseeable future.

This section begins by succinctly summarizing the current status of wind energy in global energy supply and that of 'urban wind energy', i.e. turbines placed in, on and around buildings in urban environments.

It then moves on to set wind energy within the context of the ongoing social, political and economic debates over our energy futures and sustainable development goals, which are raging at local, regional, national, international and world levels.

In industrialized nations, inefficient centralized energy supply systems (based on imported fossil fuels) are being questioned, and interest in renewable energy technologies and local generation networks is growing. Elsewhere, the need for an equitable distribution of resources to cope with environmental pressures from population growth and climate change is acute. Appropriate responses will vary, depending on the country or region being considered, however, the gravity of both issues is such as to raise concerns the world over.

The text focuses on our responses to two major challenges:

- energy security and rising energy prices; and
- environmental issues.

It discusses how wind energy (including urban installations) can and will form an increasingly important part of our energy futures.

WIND ENERGY AND GLOBAL ENERGY SUPPLY

Urban wind energy is incredibly diverse, ranging in scale from small individual wind turbines on houses to wind farms containing giant turbines on derelict industrial sites, and much more besides. It follows the spirit of the early windmills – used to grind grain and pump water – in harnessing the wind for useful work to meet local needs, i.e. bringing energy production back close to people.

Wind energy is also an increasingly important factor in global energy supply. Economic and population growth continues to lead to rapid increases in worldwide energy consumption. According to International Energy Agency (IEA) figures,[1] between 1973 and 2006 total primary energy supply for all uses almost doubled and electricity generation more than tripled (the vast majority coming from fossil fuels – coal, gas and oil). By the end of 2008, wind energy generation accounted for around 1.25 per cent of global electricity demand (installed capacity of ~120GW and electricity generation of ~250TWh).[2] This is expected to continue to grow.

Even assuming a conservative rate of growth (based on IEA figures),[3] wind energy production is expected to account for around 4.2–5.8 per cent of global demand by 2050 (depending on economic growth forecasts and the level of improvement in energy efficiency), but could reach over 20 per cent according to advanced scenarios proposed by the Global Wind Energy Council (GWEC).[4]

A high level of penetration has already been achieved in European countries such as Denmark (20 per cent), Spain (10 per cent), Portugal (12 per cent) and Germany (8 per cent), where strategic importance has been attached to the development of wind energy in national energy policies including setting of subsidies to support market growth and support at international level, e.g. by the European Union (EU).[5,6] Within the EU as a whole, wind energy generation currently meets approximately 4.2 per cent of electricity demand and saves an estimated 100 million tonnes of CO_2 per year.[7]

Global total installed capacity grew by 36 per cent in 2008 alone led by strong development in North America, Europe and Asia. The USA eclipsed Germany as the country with the largest installed capacity and the Chinese market again expanded strongly (with a doubling of installed capacity).[8]

Overall, the global market in 2008 was estimated to be worth around €36.5 billion and responsible for around 400,000 jobs.[9]

While the impacts of the global recession also began to bite in late 2008, this is expected to be no more that a temporary blip in the rapid growth of the wind energy industry.

Figure 1.1 Global electricity generation by fuel in 1973 and 2006 (IEA Key World Energy Statistics 2008)

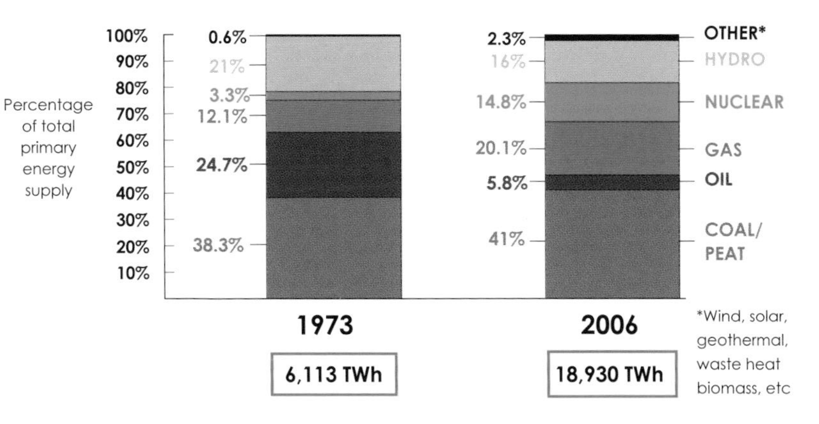

The global market for small wind (turbines) grew even faster, by over 50 per cent in 2008 according to AWEA (American Wind Energy Association),[10] with almost 40MW of turbines rated at 100kW or lower coming onstream. Of these, 28MW were rated at less than 50kW (the typical limit for classifying as 'small wind'). Small wind accounted for just over 0.1 per cent of global growth in installed wind energy capacity in 2008 (~27GW), but a distinct small wind market has emerged over the past few years and is attracting attention particularly in the largest markets – the US and UK.[11,12] In the US, urban and (building) rooftop small wind systems only account for a few per cent of this market at present, while in the UK over 10,000 systems were installed between 2005 and 2008 alone with rooftop systems now accounting for around 20 per cent of new installations.

However, AWEA predicts that a recent long-term financial incentive introduced by the federal government could increase the size of the US market by as much as 30-fold over the next five years.[13] In the UK, according to BWEA (British Wind Energy Association), the small wind industry is already responsible for almost 1900 jobs and over 50 per cent of production is exported to over 100 countries worldwide.[14]

While the small wind sector remains very small in global terms, it is likely to grow rapidly for the foreseeable future. The level of this growth depends on the key drivers. These are considered in the next section.

1) RISING ENERGY PRICES AND ENERGY SECURITY

Overview

Rising energy prices and energy security, i.e. guaranteeing national energy supplies, are the main political and economic drivers for renewable technologies such as wind energy.

Oil prices have risen significantly since the turn of the century from a relatively stable US$20/barrel (since the peak in the mid-1980s) as shown in Figure 1.2. Although prices have dropped following recent global economic events, they are expected to bounce back in the near future.

Increases in oil prices have a knock-on effect on substitutes, most notably natural gas, and both householders and industrial users have been subjected to significant additional energy costs. For example, in the EU-27 average gas price increases were 18 per cent and 17 per cent for households and 35 per cent and 9 per cent for industrial consumers from January 2005 to 2006 and January 2006 to 2007 respectively.[15] As gas- and oil-fired stations provide around one third of the EU's electricity, the electricity prices have also seen significant increases. For example, the average electricity price increases were 10 per cent and 5 per cent for households and 12 per cent and 9 per cent for industrial consumers from January 2005 to 2006 and January 2006 to 2007 respectively.

Figure 1.2 Brent Crude oil prices (Energy Information Administration, US)[16]

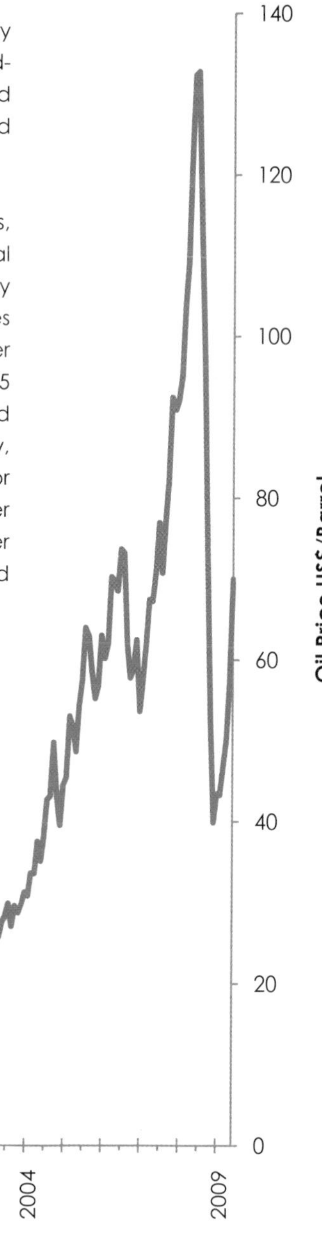

Region	Production rates (million barrels/day)	Proved reserves (billion barrels)	Lifetime at current production rate (years)
North America	13.6	59.5	11.8
South and Central America	7.0	103.5	40.7
Africa	9.8	114.3	31.8
Western Europe (DK, IT, UK, NO)	5.3	15.7	8.2
Central & Eastern Europe, Eurasia	12.3	124.8	27.9
Middle East	25.1	742.7	81.0
Asia Pacific	8.0	40.2	13.8
World	81.1	1200.7	40.6

Table 1.1 Distribution of oil reserves (BP Statistical Review of World Energy - end 2005)

At the end of 2005, the proved world oil reserves were 1200 billion barrels.[17] This equates to approximately 40 years' supply (at the production rates quoted at the end of 2005), although this date will tend to slip as more oil reserves are discovered. The proved oil reserves, however, are not the main reason for the present concerns and the associated price increases. These are attributable to two factors. First, the increase in global energy demand which requires an increase in production rates, and second (and most notably), the 'distribution' of the remaining reserves.

Increased production rates will tend to shorten the lifetime of oil reserves. Furthermore, these increased 'required production rates' are predicted to overtake 'physically possible production rate' capacity. This so-called 'peak oil' phenomenon has been predicted to occur around 2010. The extent of the 'distribution' disproportionality is shown in Table 1.1. For example, the Western European reserves (UK, DK, IT, NO) and those of the Middle East are around 10 years and 80 years respectively.

It is the distribution of oil that gives rise to energy security concerns in many countries. For example, in 2004/2005 the UK became a net importer of gas and in 2010 will be a net importer of oil (and will import 80 per cent of its gas). Gas imports to many countries come from politically unstable areas such as the Middle East, North Africa and Russia (which cut gas supplies to the Ukraine in 2006 and 2009 after a refusal to pay a substantial price increase). The existing gas import pipeline network also poses complications. It is owned and run by a variety of national monopolies or semi-monopolies on the continent and uses very long pipelines which have risks relating to physical failure or perhaps terrorist activity.

If gas supplies are interrupted the problem would be exacerbated by general low gas storage capacities. The storage capacity of countries such as France, Germany and Italy is around 20 per cent of the annual demand while some countries have a very low storage rate, such as the UK, which has a storage capacity that amounts to two weeks supply or 4 per cent of the annual demand.[18]

Wider political issues relating to rising energy prices include the effect on national inflation levels, freight costs and borrowing, and can lead to investors and companies favouring to site their businesses in countries which do not have energy issues.

Considering just these basic ideas, the upward trend of future energy prices can be understood. Consequently, the increasing force of these drivers to find new solutions to the energy supply situation can be appreciated.

What role should nuclear energy play in the future of energy supply?

The fuel for standard 'once through' reactors has been quoted to have a proved 'reserve life' of the same order as fossil fuels and therefore the current nuclear power technology is, at best, a means to buy time to develop other technologies. However, resource prospecting is at a much lower level than that of the oil/gas industry and the actual cost-effective nuclear fuel reserve life is likely to be much higher.

However, this abundant energy resource does come with some serious implications. Public and environmental protection is one of the major concerns. The permanent safe storage of radioactive waste still remains an unsolved problem and the long-term effects of radiation exposure are not fully understood. Waste can remain dangerous, and has to be reliably stored, for hundreds to thousands of years for some plutonium isotopes and other materials. Ensuring the safe operation and correct decommissioning at the end of the 40-year design life comes at considerable expense. Even in an experienced developed country, reactors have proved difficult to manage financially despite extensive government assistance. Operating nuclear plants in less developed countries may cause complications if, for example, economies become impoverished during the operation life of a nuclear facility.

Figure 1.3 Wind energy existing alongside a nuclear power station at Pickering on Lake Ontario, Canada. A Vestas 1.8MW turbine with a blade diameter of 78m which is 150m from the nearest occupied buildings and reported to meet the electricity demands for 600 homes. (Harold L. Potts)

Although nuclear plant operation is often quoted as having no direct emissions of gases such as carbon dioxide (CO_2), sulphur oxides (SO_x) and nitrogen oxides (NO_x), other unnatural gases are released and 'indirect' gaseous emissions can be significant as discussed later.

Are decentralized energy networks viable?

Renewable energy technologies, such as wind energy, naturally lend themselves to decentralized energy systems (where energy is generated in networked distributed nodes local to energy users). But can decentralized networks be viable and provide the same quantities of energy as conventional centralized power plants?

Perhaps the focus should not be on how to get more energy sources but how to fit the right energy supply to the energy end use in question. Some proponents of decentralized and distributed energy systems (e.g. Amory B. Lovins) have detailed how today's centralized electricity generation industry (large power stations feeding a nationwide grid) will become a thing of the past.[19] Lovins's 'soft energy path' proposes the use of diverse production methods matched specifically to end uses in scale and quality, coupled with energy-efficient systems.

The necessary diverse energy portfolio would not only include wind turbines, but solar energy in several forms, energy from biomass and waste, geothermal energy, and can even extend into high-efficiency gas combined heat and power (CHP) plants.

Although latest generation CHP units are typically gas-fired, they are compact, flexible and efficient. The flexibility means that they can pick up shortfalls in renewable energy production (e.g. wind/solar have varied production profiles). Standard gas-fired CHP transforms 23–28 per cent of the energy content of the fuel into electricity, 55 per cent into high-grade heat (70–80°C) and 10 per cent as lower-grade heat (40–50°C) with less than 10 per cent lost through the flue and case (i.e. room heating). This heat can be used for both heating and cooling (via absorption chillers). Standard large-scale gas-fired power stations convert around 40 per cent of the energy content of the fuel into electricity with the remaining 60 per cent as heat losses. These CHP/renewable networks are scalable (providing relatively easy means of growth and network expansion) and can be embedded in regional distribution networks rather than exporting to the grid.

The sustainable community energy system at Woking, UK, is an example of how this type of new thinking can not only be practical but also profitable (see Figure 1.4). The initiative was led by Allan Jones who when interviewed said: 'To be truly sustainable, you have to go back to communities. In Woking, we had 60 "island generation sites" [independent energy-producing areas] and I think the term island is the right one because it is a case of building a little barrier around yourself so you are making yourself self-sufficient.'[20]

As an aside, it should be noted that when energy networks are being designed, energy quality (exergy) should be kept firmly in mind in order to make the most of energy resources and minimize 'waste energy'.

How much energy can a turbine generate?

Energy output from turbines can be a difficult subject to approach without some prior understanding of the basic principles. In the commercial world, there are several terms to negotiate around including 'turbine rating', capacity factor', 'installed capacity', 'annual energy output' and 'capacity credit'. Box 1.1 outlines the important relationship between these terms ('capacity credit' is expanded on below when the reliability of wind energy is discussed).

To further complicate the situation, the discrepancies between manufacturers' predicted yield and the output from the final installation have, in certain instances, been disappointingly large. In these cases it can be difficult to know whether this is due to inadequate local wind resources, a manufacturer artificially inflating figures or whether the turbine has been stalled in an inappropriate manner.

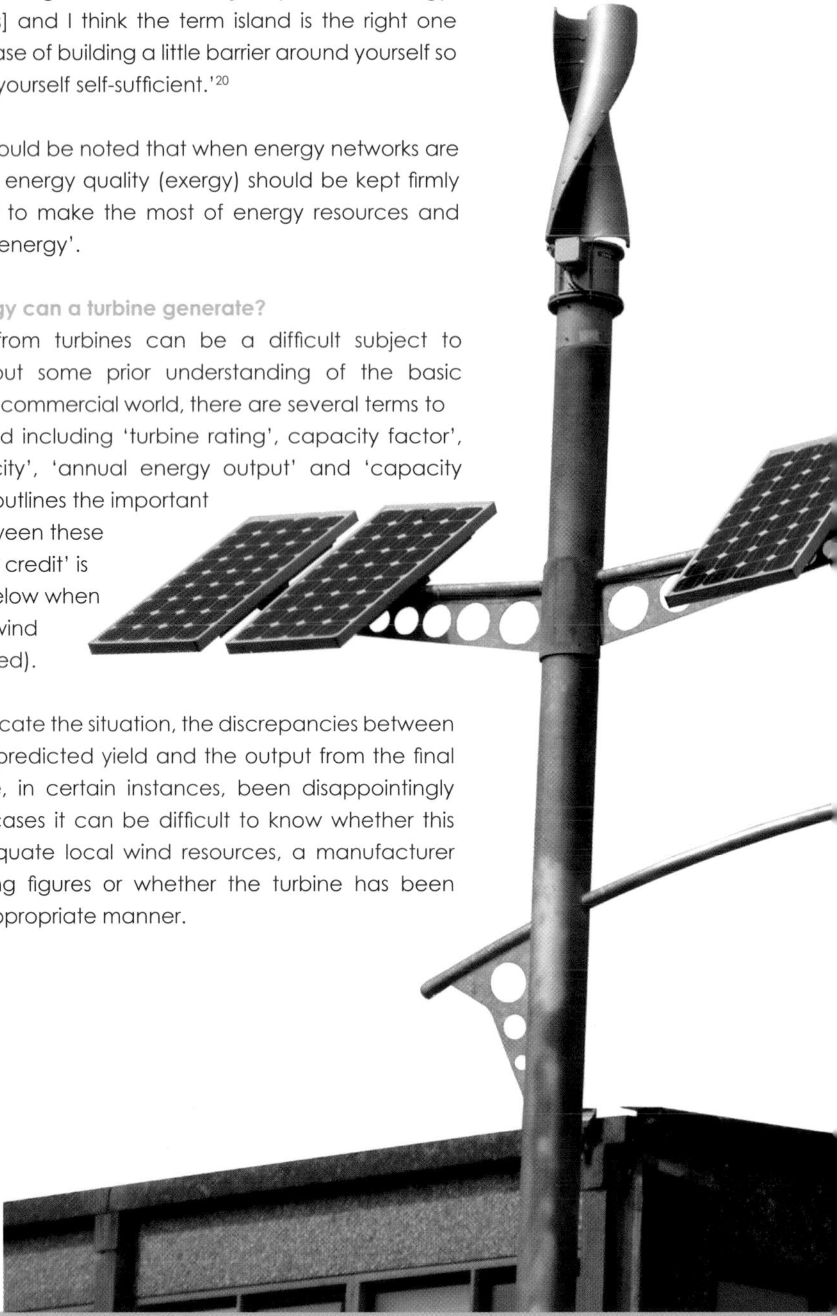

Figure 1.4 Vertical axis wind turbines (VAWTs) and photovoltaics (PV) as part of the sustainable community energy system at Woking, UK (Susana Espino)

The most reliable way to predict the energy output from a proposed installation is to go back to basics. This involves calculating the actual 'available energy' in the wind based on the 'swept area' of the blades and using reliable local site wind data. The actual annual energy production from a turbine can be estimated by multiplying the available energy in the wind with the turbine 'coefficient of performance' which is taken from physical testing data (ideally over a full range of wind speeds).

BOX 1.1

TURBINE RATING, ANNUAL ENERGY OUTPUT, CAPACITY FACTOR, AND INSTALLED CAPACITY

When discussing the output from wind turbines both 'power' and 'energy' are used. Power can be thought of simply as the instantaneous capacity to do useful work and is given in watts (joules per second) or kilowatts (1kW = 1000W). Energy is total capacity to do useful work over a given period of time (joules) and is commonly given in kWh which corresponds to 1kW produced (or consumed) for one hour (this is equal to 3600KJ).

The 'rating' (or nameplate capacity) a turbine is given by a manufacturer is the maximum 'power' a turbine will produce. This peak output usually occurs when the wind speed is around 12m/s. Typical outputs will be much lower (e.g. 20 per cent of the rated value) at average wind speeds (of around 6m/s). The speed at which different manufacturers rate their turbines can vary (usually between 10 and 13m/s) making direct comparisons more difficult.

The 'annual output of turbine' (given in kWh) can be a more useful parameter to consider. It can help to determine the relevance of a certain turbine in relation to a particular electricity demand and to evaluate associated income. Again, values quoted by manufacturers can be difficult to compare directly as the number of kWh a turbine is reported to produce (in ideal conditions) will depend on the average mean wind speeds used for the calculation and the distribution of wind speed frequencies.

The 'capacity factor' is the ratio of the actual output of a turbine over a period of time and the output if that turbine had operated at the full nameplate capacity over the same period. It is a simple measure of the overall average performance and takes into account factors such as the local wind speeds, turbine efficiency and the availability. Typical capacity factors are often quoted to be around 30 per cent for large-scale wind farms in areas with good wind resources although they can be closer to 20 per cent depending on the local conditions. Note that the capacity factor is not the same as 'coefficient of performance', which is also usually taken to be around 30 per cent.

The 'installed capacity' is simply the sum of the turbines' rating. For example, if 10 turbines rated at 2MW are in operation in a certain location then the installed capacity is 20MW. If the capacity factor of a given site was 20 per cent, the averaged instantaneous power would be 4MW. This would give an 'annual energy output' of 35,000MWh/a and meet the electricity demands of 8760 gas-heated three-bed homes (or the total energy demands of the same amount of homes meeting the Passive House standard).

Using this approach, an energy production summary for different sizes of generic horizontal axis wind turbines (HAWTs) is given below in Table 1.2. It should be noted that this summary assumes an average turbine coefficient of performance of 0.3 and a 'Rayleigh distribution' of wind speeds (these terms are expanded on in Part 3). This simplified and generalized table shows that these devices are capable of generating significant amounts of energy. The table also highlights the importance of the size of the turbine. It should be noted that a well-designed and well-positioned turbine will produce more energy than quoted in Table 1.2 (provided the annual mean wind speed is at least 5.5m/s at hub height). The importance of the quality of the available wind resources (mean wind speed) is reflected in the energy factors given in Table 1.3 (which are relative to an annual mean wind speed of 5.5m/s) for local site mean wind speeds from 4m/s to 9m/s. For example, moving a turbine from a site that has a mean wind speed of 5.5m/s to one that has a mean wind speed of 7m/s will more than double the energy output.

Table 1.2 Typical available energy and supply potential (number of homes) from a range of horizontal axis wind turbines (HAWTs) when annual mean wind speeds are 5.5m/s at hub height

HAWT blade diameter	Blade swept area	Energy capture			Potential use	
		Mean wind speed = 5.5 m/s			Standard home @ 4000kWh/a	Passive House @ 1500kWh/a
		Power In wind	Power from turbine	Annual energy		
(m)	(m^2)	(kW)	(kW)	(kWh)	(No. of homes)	(No. of homes)
1	0.8	0.1	0.02	374	0.09	0.25
2	3.1	0.3	0.09	1496	0.37	1
5	19.6	2	0.56	9350	2	6
10	78.6	7.8	2.24	37,401	9	25
15	176.8	17.6	5.03	84,153	21	56
20	314.3	31.4	8.94	149,605	37	100
25	491.1	49	13.97	233,758	58	156
30	707.1	70.6	20.12	336,611	84	224
35	962.5	96.1	27.38	458,166	115	305
40	1257.1	125.5	35.77	598,420	150	399
50	1964.3	196.1	55.88	935,032	234	623
60	2828.6	282.4	80.47	1,346,446	337	898
70	3850	384.3	109.53	1,832,662	458	1222
80	5028.6	502	143.06	2,393,681	598	1596

A constant overall turbine 'coefficient of performance' of 0.3 has been assumed to calculate 'power from turbine'. A 'standard home' approximates to a three-bed gas-heated house in the UK with an annual electric demand of 4000kWh/a. This is roughly equivalent of the total energy requirement from a three-bed 100m² 'Passive House' (Passiv Haus).[21] A Passive House has a 15kWh/m²a energy requirement for space heating and the total energy consumption of a Passive House was taken to be 42 kWh/m²a by the CEPHEUS project,[22] which includes space heating, domestic hot water and household appliances.

Site mean wind speed (m/s)	4	4.5	5	5.5	6	6.5	7	7.5	8	8.5	9
Energy factor	0.38	0.55	0.75	1.00	1.30	1.65	2.06	2.54	3.08	3.69	4.38

Table 1.3 Energy factors (relative to a site mean wind speed of 5.5m/s) for local site mean wind speeds

Can the wind resource be relied upon for a secure energy source?

The two main concerns voiced in relation to wind energy are the fluctuating nature of wind and the amount of conventional generation it can displace. From the point of view of the conventional centralized grid operators, wind energy has been assigned a low 'capacity credit'. This is a measure of confidence for guaranteed energy supply. Other renewable energy sources such as hydro, tidal, geothermal and biomass, in contrast, have a much higher capacity credit.

If only one turbine was connected to the grid the capacity credit would be zero as no electricity is produced when the wind is below say 4m/s, i.e. there is no guaranteed supply. However, when several turbines are distributed over a wide area the capacity credit increases. This is due to the fact that the wind may still blow in one location even though conditions may be calm in another area.

A typical wind capacity credit may increase to 10 per cent of the 'installed capacity' (i.e. around half of the 'capacity factor'). However, there is a point where an increase in turbine density causes the capacity credit to decrease. For example, in Germany, with its relatively high installed capacity of 17,000MW (2006), the capacity credit was 8 per cent. This is predicted to fall to 4 per cent if the installed capacity was to increase to 48,000MW. This would replace 2000 MW of guaranteed capacity, i.e. 24,000 2MW wind turbines would replace two conventional medium-sized coal stations.[23] It has been suggested that wind energy can cause power stations to run at a reduced capacity and therefore at much lower efficiencies (thus wasting energy). However, power networks already manage extremely variable loads due to varied demand.

The ability to manage the fluctuating nature of the wind can come from both ends of the energy supply chain. From the supply end, increasingly sophisticated wind prediction techniques are being used and developed. From the distribution end, a more effective energy-generating portfolio, refined local distribution networks and even energy storage (e.g. using hydrogen, via pumping and damming of water, or through electrochemical means such as VRB (vanadium redox battery[24]) offer potential solution paths. Furthermore, active 'energy demand management' techniques can be used. These offer capital incentives for end users to manage their own energy consumption to coincide with periods of low demands and high supply.

The typical maximum penetration of wind energy has been quoted as 20 per cent. However, this implies the continued use of existing grid systems. If appropriate systems are allowed to evolve, as pointed to above, then this limit could be comfortably exceeded. Interestingly, the existing capacity of hydroelectric power plants in Europe could store enough energy to meet European electricity demand for one month.[25] This type of storage technology could be used to meet peaks in supply. However, greater inefficiencies exist with increased conversion of energy through different forms, e.g. moving energy from wind electricity to mechanical pumping, to gravitational potential energy and back to electrical energy is less efficient than using electricity directly from wind energy. Physical and political development of international cooperating electricity grids may be one route to allow these types of systems to be achievable.

In the long term, of course, wind energy and other renewable energy technologies are the only energy sources that can be relied upon.

Are wind turbines reliable?

The reliability of wind turbines is referred to as 'availability' and is a measure of the percentage of the year a turbine is available to generate electricity. The availability of large-scale wind turbines has improved considerably over the last two decades as the technology has developed and experience levels have increased. Today's 'mega-watt' turbines have an availability of more than 97 per cent. This translates to an 'off-line' period for maintenance occurring for less than 3 per cent of the year (i.e. about a week). This development is not surprising given the large sums of capital investment that have already been made in this industry.

Some smaller-scale wind turbine manufacturers are relatively new to the market, and in some cases the companies are very small. Although there is some knowledge transference down from large-scale development, the technology can be significantly different. Also, experience and investment levels are much lower with small-scale turbines. However, there are a number of fairly well-established small wind manufacturers attempting to progress the field in this important early development stage (see Appendix 1 for a list of manufacturers). For smaller turbines located in urban areas the reliability issue holds significant importance. An extensive maintenance requirement would not only have financial repercussions but negative implications for public perception and future investment levels.

Due to the size of these companies and the limited long-term experience with the technology, the amount of information made available on the performance of their turbines can also be limited. It is clear that manufacturers able to invest significantly in the development of their product (and have others willing to use their product) will emerge as leaders.

As wind turbines have moving parts the product lifetimes will of course be finite. However, design lives are often quoted to be around 25 years and certain manufacturers offer fairly comprehensive long-term product guarantees. Re-engineering, i.e. the replacing of components at the end of their design life, is becoming a more established practice in large-scale wind.

Can wind turbines be economical?

Wind turbines can be economical if enough wind resources are available (a rule of thumb is an average wind speed at hub height of 5.5m/s or more). Large-scale turbine wind farms are becoming more and more widespread and the number of planning applications for new wind farms indicates high levels of enthusiasm for these projects by developers. This is in no small part due to their current economic profitability (which can be boosted by subsidies). However, transmission cost can be high. For example, the National Grid (UK) has estimated the cost of grid expansion to accommodate a proposal for very large scale wind installations in regions of north-west Scotland far from urban areas (the windiest part of the UK) to cost £250,000 per MW, which is more than the cost of the turbines themselves.

With large-scale urban wind installations, which usually involve installing one or small groups of turbines, the economies of scale usually found with wind farms are not present. However, there are cost benefits associated with urban installations which can offset the differences, e.g. lower distribution costs, lower access road costs and reduced foundation costs.

The decrease in energy output from having local obstacles restricting the wind flow can be recouped by the low transmission losses of urban turbines (typical UK grid transmission losses are around 7 per cent). Also having the opportunity to supply electricity directly to the end user/customer can increase profitability significantly. In most countries, this option is preferable to exporting directly to the grid, although some regions have generous 'feed-in tariffs' set by the government specifically to act as an incentive for wind energy, or in some cases the power companies themselves may choose to offer high feed-in rates for wind energy. The UK has opted for Renewable Obligation Certificates (ROCs) and initial capital cost grants to compensate for the typical low feed-in rates offered by most (but not all) utility companies. This interest scheme is discussed further in Part 3 and in Box 3.4.

This claim that urban wind energy can be economical is perhaps best demonstrated by going to first principles and considering the numbers for a simple large-scale installation case - see Box 1.2. A criticism of wind farms often comes from the possible negative economical impact through loss of rural tourism (e.g. a £14billion-per-year industry in England supporting an estimated 800,000 jobs). In general, wind turbines, and the associated additional pylons to transport the electricity from remote areas, detract from the enjoyment provided by rural areas. Many local businesses suffer if tourism decreases. A survey in rural Scotland found that some 10 per cent of visitors may not return to an area if a wind farm was built.[26]

However, a community-owned urban turbine in a rural town could certainly provide a permanent 'financial well' where the profits can contribute to enhancing the surroundings, adding to the welfare of the community and therefore boosting tourism. One may anticipate that rising costs of energy and the energy security issue may tend to increase acceptance of wind energy. However, over-exploitation or inappropriate exploitation of wind energy would increase any dissatisfaction. Wind energy is certainly not a suitable sustainable solution in many areas and should not be shoe-horned into inappropriate locations. Sustainability by definition incorporates economical as well as wider social and environmental concerns, and planning committees are in place to regulate their installation in the most effective manner. All wind energy proposals should be judged on a case-by-case 'first principles' basis while keeping the wider picture in focus.[27]

The future will see economic viability tending to improve as fossil fuel energy prices rise and the cost per kW of installed wind energy decreases (while the performance of newer designs increases).

BOX 1.2

SIMPLE FINANCIAL MODEL FOR WIND ENERGY VIABILITY

Indicative finanicial figures for one free-standing 70m blade-diameter turbine sited near a community where average wind speeds are 5.5m/s at 60m above ground level are:

- Typical installation costs £1.0 million.
- Energy in the wind 384kW for the given swept area based on average annual wind speeds of 5.5m/s.
- Energy capture by the turbine (coefficient of performance of 30 per cent) is 110kW.
- Energy captured per year (including downtime for service) 1,800,000kWh (Rayleigh wind speed distribution) i.e. enough electricity for over 450 standard three-bed homes.
- Price paid per kW by domestic customer 10p/kW.
- Over ten years the revenue equals £1.8 million.

This can be conservative as:

- the energy in the wind is greater at higher wind speeds (as discussed later) and at 6m/s at hub height the revenue over ten years would be £2.3 million;
- the average coefficient of performance over the key wind speeds is greater than 30 per cent for modern turbines (even taking transmission losses into account);
- price paid per kW by domestic customer can be more than 10p/kW. Energy prices tend to rise and customers may be willing to pay more for 'green electricity' that may benefit the community directly and visibly;
- schemes such as grants or ROCs have not been taken into account (which provide an extra 4p/kWh);
- turbines can be expected to last for 20 years (and can be re-engineered at the end of their design life);
- initial capital costs can be decreased considerably with 're-engineered' (second-hand) turbines.

A more detailed economic assessment would take into account additional factors such as the cost of maintenance, distribution losses, interest on capital, taxation and inflation. However, on the whole it is clear that wind energy can be viable and profitable for even modest wind resources (5.5m/s at hub height).

2) ENVIRONMENTAL ISSUES

What are the impacts of climate change?

The Earth's climate is a hugely complex system dependent on many other complex systems. A growing body of evidence suggests this system – our climate – may be changing at rates beyond natural fluctuations. Direct effects of climate change include the occurrence of more frequent and extreme weather and rising sea level resulting from melting ice-caps and the expansion of the sea. The consequence of these effects from human welfare, economic and biodiversity points of view are far-reaching to say the very least. Furthermore, the consequence of changes in one aspect of this ecosystem can cause more serious knock-on events.

It is difficult to know which of the emerging theories are credible. Some range from situations where the reverse of global warming is achieved (i.e. the initiation of an ice age) through disruption of the Gulf Stream, to the initiation of a series of runaway chain reactions. These could include the release of huge quantities of dissolved methane (CH_4, a potent greenhouse gas) from stores in the sea (via an the increase of sea temperatures), or methane releases from thawing lands in Siberia, or increase in CO_2 releases through destruction of green areas which result from adverse weather conditions or forest fires. Similarly, the reduction of polar ice, which reflects some of the sun's short-wave radiation away from the Earth, would also speed up climate change.

Should these theories represent reality the potential seriousness is clear. The cost of inaction may be huge. Figures have been suggested, e.g. US$5 trillion over the next century,[28] to account for attempts to prevent and then treat a number of potential outcomes such as:

- direct humanitarian disasters (e.g. via storms, short-term flooding, drought, famine);
- indirect humanitarian impacts (e.g. damage to agriculture, forestry, fisheries and resource distribution networks, infrastructure, pollution);
- degradation of the planet (e.g. loss of land, woodlands, wetlands, coastlines and species);
- associated human responses (e.g. effect on economic stability and investments, migration and conflict);
- degradation of health and healthcare.

Climate change therefore could potentially have a major impact on the future planetary wealth. Climate does of course vary naturally and only 15,000 years ago half of the world was covered in ice. Therefore examining to what extent our current cultural behaviour (short term, individualistic, consumeristic) may be responsible for these potential outcomes is important.

How dependent is climate change on man-made CO_2?

Information on the potential impacts of climate change is readily available and for many awareness is already high. Initial guidelines for action to mitigate climate change have been presented in many areas but implementation seems to be slow. Perhaps one of the main reasons for the slow uptake in establishing new practices, aside from economic issues, is the uncertainties related to the efficacy of the proposed mitigation methods.

The theory gaining the widest acceptance relates to carbon emissions. CO_2 released from the processing of underground stores of fossil fuels, together with large-scale deforestation, is thought to increase the strength of the 'greenhouse effect' (a phenomenon known to allow short-wave solar radiation to penetrate the atmosphere and warm the Earth's surface while trapping long-wave radiation attempting to leave the planet). However, the situation is rather complex.

There are many sources of atmospheric CO_2 emissions, among which the anthropogenic emissions are a small percentage. However, there are many natural carbon sinks, which have until the last century maintained an approximate balance with the natural sources.

There are of course many factors other than CO_2 atmospheric concentration which regulate the temperature of the land, sea and the atmosphere. For example, both water vapour and cloud formation have a much more significant blanketing effect than CO_2. However, water vapour is not generally thought of as a 'forcing' variable (i.e. forcing the climate change in any particular direction) as its value tends to stabilize in short timescales relating to humidity levels. However, some theories suggest water may have a 'forcing' action on the climate through, for example, aircraft emissions at very high altitudes (and this applies to both H_2O and CO_2).

The extent of the 'blanketing effect' of cloud cover on long-wave radiation transfer is considerable. Its effect is perhaps more commonly felt at night-time, when long-wave radiation exchange from the skin can be felt to increase cooling during the absence of cloud cover. Unfortunately, long-term historical water vapour concentration cannot be determined (e.g. from ice core samples) and so a more complete picture of the long-term history of the atmospheric composition cannot be obtained.

There have been claims that, over the last several hundreds of millions of years, CO_2 concentration increases generally preceded, and therefore caused, global temperature change. Ice core data from the Antarctic have been inconclusive although some argon isotope analysis suggests it may have been temperatures which have generally risen first (occurring some 800 +/- 200 years before CO_2 rises).[29] However, these findings do not mean that humans artificially increasing CO_2 levels will not cause global warming. The rapid release of man-made CO_2 from the ground and in the skies is unprecedented. Outcomes are therefore difficult to predict from long-term climate analysis, although early signs and present-day evidence are beginning to point to unsettling conclusions.

A number of other theories also exist to account for recent temperature changes (~0.8°C since the 1980s).[30] For example Danish scientists posed the theory that global warming could arise through variations in solar cosmic ray emissions, after carrying out laboratory experiments on increased precipitation brought about by the presence of highly energetic particles. Other theories include variations in the orbit of the Earth (Milankovitch cycles) that are known to alter levels of incident solar radiation.

The predicted levels of temperature increase from climate simulations have been higher than we have actually seen. This may be, in part, due to the promotion of organic growth a carbon-rich atmosphere provides (and increased CO_2 absorption). For example, the density of the Amazon rainforest (relating to carbon uptake) has increased in recent years. It has been claimed that the 'lungs of the Earth' have been buffering and limiting climate change (as well as a buffering from CO_2 absorbed by the sea). Another theory states that this buffering is reaching its limit and this carbon will be released should climate conditions diminish the size of the rainforest and warm the sea.

Although the focus has been on CO_2 there are many other greenhouse gases. The extent to which these types of gases absorb long-wave radiation varies significantly. Methane (CH_4), for example, is a much stronger greenhouse gas than CO_2. Also, around 70 per cent of methane emissions are estimated to be due to anthropogenic causes compared to around 4 per cent of carbon emissions. These methane releases stem from sources such as power generation, cows, rice fields and landfills/waste disposal. However, the atmospheric concentration of CH_4 is much lower than CO_2 as within a decade it decomposes into CO_2 and H_2O. The climate forcing strength of methane is estimated as around one third of that of CO_2.

Despite the complexity and the level of uncertainty, many governments and non-governmental organizations (NGOs) have demonstrated a level of confidence in the conclusions drawn by the Intergovernmental Panel on Climate Change (IPCC) on climate change and CO_2. The Kyoto Protocol (1997) put forward the initial targets to reduce CO_2 emissions by 12.5 per cent below 1990 levels by 2010. Since then EU leaders have agreed to cut carbon dioxide emissions by 20 per cent from 1990 levels by 2020. Some countries have elected to go further. For example, the UK government elected to set a target of 20 per cent below 1990 levels by 2010. A UK White Paper stated that at least half of this target would come from energy efficiency with renewable energy to account for the rest. The target of 10 per cent of total UK electricity use to come from renewable sources has been set. The UK government is also committed to reducing carbon emissions by 60 per cent by 2050 with a recommendation that 20 per cent of UK electricity should be generated from renewables by 2020.

Where are the main sources of carbon emissions?

In the UK, for example, electricity generation is responsible for one third of CO_2 emissions. The rest is derived from industrial processes and construction, vehicle exhaust, domestic heating and aircraft. Of these, the fastest-growing source of CO_2 emissions is air travel. This is expected to double by 2025.

Is nuclear energy a low carbon emitter?

Typical CO_2 emissions from the operation of standard 'once through' nuclear reactors using high-grade ores are approximately 30 per cent of the emission levels of a gas-fired electric power plant. In the long term this figure will tend to rise as the quality of the grade of fuel being used decreases and more energy is used to mine, refine and enrich the fuel. In theory, it could reach over 100 per cent of the CO_2 emissions of a gas-fired electric power plant for 0.02 per cent grade for hard ores and 0.01 per cent grade for soft ores.[31]

How much carbon does the energy from one wind turbine save?

Wind turbines reduce carbon emissions by displacing conventional electricity generation. The amount of carbon saved by a turbine therefore depends on the amount of energy it can produce. The size of the turbine (swept area) is one of the key factors in terms of energy generation and this can vary greatly, with the smallest turbines usually having a blade diameter around 1m (rated around 100W) and the largest production turbines (rated at 5MW and 6MW) having a blade diameter of 126m (Figures 1.5 and 1.6 respectively).

The amount of carbon a turbine saves also depends on the type of conventional generation being displaced. National grids are usually supplied by electricity generated from a number of fuel sources, such as gas, nuclear, coal, oil, hydro and other renewables such as wind energy. The mix varies greatly between countries and is in a constant process of evolution depending largely on the changing market conditions. The demand for electricity by the various types of end users also varies greatly, on both a seasonal and daily basis, and so the realities of grid generation are complex. Therefore, when estimating the amount of CO_2 saved per kWh produced by wind energy a defensible stance should be taken. For example, in the UK, the British Wind Energy Association (BWEA)[32] have used 0.86tCO_2/MWh in their literature. The Department for Environment, Food and Rural Affairs (Defra) use 0.43tCO_2/MWh which represents the current 'mix' of energy sources (although the Department of Trade and Industry (DTI) has used 0.65 tCO_2/MWh for their mix). By 2010, when cheaper gas displaces the current coal-fired stations, the conversion factor has been predicted to fall to 0.27tCO_2/MWh.[33]

A common point raised by those wanting to protect the countryside (and tourism) from the proliferation of wind farms is the idea that wind energy causes power stations to operate at less than their optimum output levels. A reduction in efficiency could negate any saving in CO_2 emission. However, it is always in a generation company's interest to maximize efficiencies and it is likely that only those already operating in a 'variable' mode to account for the variability in users' demand will remain in this mode.

Figure 1.5 Microturbine (Rutland WG-317) charging a barge 12/24V battery (from UK manufacturer Marlec). The turbine is rated at 90W at 10m/s and 24W at 5.1m/s with a 0.9m blade diameter. (Dr Matthew Overd)

Figure 1.6 The largest production turbine from German manufacturer RePower rated, at 5MW and with a 126m blade diameter (Jan Oelker & RePower)

Any electricity generation due to new renewable energy projects is likely to displace the most expensive form of energy generation (i.e. coal-fired station). This is the greatest emitter of CO_2 for electricity generation.

The factor selected to convert the energy generated from a wind turbine into CO_2 saving (e.g. 0.43tCO_2/MWh) will not include the carbon emissions associated with design, manufacturing, transport, foundations and access roads, power cables and substations, operation and maintenance, and decommissioning. However, whether the full extent of CO_2 emissions for fossil fuels generation has been fully accounted for in the figures given above is unclear.

The amount of carbon a turbine saves also depends on how the project has been designed as well as the lifetime/reliability of the turbine. The turbine will have to run for a given period of time just to repay the CO_2 related to the embodied energy (energy related to manufacturing, transport, installation and decommissioning not accounted for in Table 1.4). Generally for larger turbines carbon payback can be around 3–12 months[34] depending on the amount of development ancillaries such

Table 1.4 Typical available energy and corresponding CO_2 savings from a range of horizontal axis wind turbines (HAWTs) when annual mean wind speeds are 5.5m/s at turbine hub height

HAWT blade diameter (m)	**Blade swept area** (m²)	**Energy capture**			**Carbon savings**		
		Mean wind speed 5.5 m/s			Coal-fired power station tonnes CO_2/year	Current UK generation mix tonnes CO_2/year	Gas-fired power station tonnes CO_2/year
		Power in wind (kW)	Power from turbine (kW)	Annual turbine energy (kWh)			
1	0.8	0.1	0.02	374	0.4	0.2	0.1
2	3.1	0.3	0.09	1496	1.5	0.6	0.4
5	19.6	2.0	0.56	9350	9	4	3
10	78.6	7.8	2.24	37,401	37	16	10
15	176.8	17.6	5.03	84,153	82	36	23
20	314.3	31.4	8.94	149,605	147	64	40
25	491.1	49.0	13.97	233,758	229	101	63
30	707.1	70.6	20.12	336,611	330	145	91
35	962.5	96.1	27.38	458,166	449	197	124
40	1257.1	125.5	35.77	598,420	586	257	162
50	1964.3	196.1	55.88	935,032	916	402	252
60	2828.6	282.4	80.47	1,346,446	1320	579	364
70	3850.0	384.3	109.53	1,832,662	1796	788	495
80	5028.6	502.0	143.06	2,393,681	2346	1029	646

as distribution lines, access roads and foundation. For smaller turbines this figure is more difficult to generalize due to variation in design; however, it is likely to be around a year. HAWTs have a lower embodied energy than VAWTs as the amount of material per m^2 of swept area is much less.

The total 'tonnes of CO_2' saved by a given turbine can be used to compare with other activities. For example, the CO_2 saving from one well positioned 2m diameter HAWT run for one year are negated by one single short-distance (4 hours) return flight.

Generally, the idea of 'tonnes of CO_2 saved' by using any form of renewable energy is simply a means to present something tangible to inspire action and whether, for example, 0.43tCO_2/MWh or 0.27tCO_2/MWh is used in the calculation is largely irrelevant. The main conclusion is that wind energy does not release CO_2 into the atmosphere, while fossil fuels and nuclear energy do.

Are the impacts of turbines too small to be considered worthwhile?

It has been suggested that the impact of wind turbines on CO_2 emissions (and other pollutants such as NOx and SOx) is too small to be worthwhile. These claims have some validity in 'simplistic' quantifiable terms. The carbon savings from one turbine generating electricity will have a negligible positive influence on climate change.

This idea can be extrapolated to a national level. For example, achieving the 2010 UK renewable energy target of 10 per cent of renewable energy generation would reduce about 2 million tonnes of CO_2 each year. This is only a small percentage of total UK emissions (550 million tonnes). Furthermore, the UK is estimated to be responsible for only 1.5 per cent of the total anthropogenic CO_2 emissions, and so it may seem that contributions from the whole of the UK's renewable energy efforts cannot make a significant contribution. Extrapolated to a global level, even if Kyoto Agreement targets are reached by 2010, predictions have suggested that these worldwide efforts will make only a minor difference in curtailing predicted temperature rises by 2100.

However, in 'systemic' terms statements on the ineffective nature of renewable energies hold less validity. In complex human systems the influence of seemingly small events can have very widespread impacts on other events, which influence further impacts in a synergistic manner.

Positive actions in one area encourage those in other areas. They raise awareness, demonstrate what can be done, demonstrate accountability (provide role models), provide experience, add investment into the area (in terms of technology and human skills) and generally shift the inertia of cultural thinking and behaviour.

The Kyoto target is a relatively small step. However, it is only the first step on a 'ladder of change'. There are of course natural limits to any first step which can only be so high before preventing any action at all. Once humanity begins to move its thinking and the direction of its actions it may be possible to counteract a number of issues such as climate change one decision, one action, one project, one step at a time.

On reflection, this first step – the Kyoto target – may appear quite substantial for a largely consumer-based culture. It may be possible for a second step to be even greater (as more countries join and the new targets/directions/outcomes are based on valuable learning from the previous step). Further progress can be built on each step. Systems often demonstrate 'tipping points' where the 'snowball' begins to roll and gather speed and size without having to expend effort pushing (i.e. concentrating on the 'doing' rather than simply 'raising awareness').

Shouldn't money be spent on other things?

Wind energy is one of the cheapest renewable electricity generation technologies. However, with only limited capital resources other areas vie for their share. Many of these other areas can be thought of as more deserving. These are sometimes referred to as 'low-hanging fruit', i.e. areas which reduce carbon emissions in a relatively simple, cheap and effective manner. One such example is improving energy efficiency by moving from incandescent lighting to compact fluorescent lights.

Many energy efficiency measures can be relatively cheap and simple to implement and, as a rule of thumb, a degree of energy efficiency should be put before renewable energy generation. The Intergovernmental Panel on Climate Change (IPCC), for example, advocate general energy efficiency in the areas of building, manufacturing and transport over wind power.

Other areas where resources can perhaps be better spent include energy-efficient generation, carbon sequestration and direct carbon capture techniques (see Box 1.3). Renewable energy technologies, like carbon sequestration techniques, require development. However, one may be inclined to view renewable energy investments as a more worthwhile medium- to long-term investment rather than investing in fossil fuel systems and the mitigation of their emissions. Choosing to invest time and money into the renewable sector will also encourage similar action from others (e.g. manufacturers will be able to invest further in their products). It may seem sensible to want to develop new systems rather than developing pseudo-fixes for inherently poor systems.

It should be noted that, currently, developed countries are poor role models for aspiring developing countries in terms of energy. New thinking in the area of energy networks may serve to provide suitable models for emerging countries to follow.

Another use of financial resources is preparing to handle what may seem like 'inevitable' consequences. This may make sense if implementing a cure is more expensive than the cost of treating the consequences. Unfortunately, negative consequences resulting from climate change will have the greatest impact on developing countries which have less capacity for adaptation. A huge investment in present-day developing countries, on the same scale as climate change mitigation, would provide infrastructure and resources to cope with future change.

In this regard, decentralized energy generation techniques and expertise may be better for providing some developing countries with energy. These 'energy nodes' can be developed at the same rate as the demand grows.

If climate change is seen as a runaway train then it can seem that efforts to try to 'slow the train down a little' are less well spent than preparing the town to deal with the impact. The difficulty lies with the fact that the impact of this 'train' is continuous and perhaps increasing in magnitude. The Stern Report review of the economics of climate change commented that the cost of tackling disruption to people and economies would be between 5 and 20 per cent of the world's output compared to 1 per cent to stop and reverse the effects.

BOX 1.3

CARBON CAPTURE

'Carbon capture' is sometimes called 'carbon sequestration' or 'CCS' (carbon capture and storage) and refers to measures taken to directly extract and store CO_2 which would otherwise contribute towards CO_2 levels in the atmosphere. It can be thought of as one of the three main areas which have been conceived to deliberately attempt to lower CO_2 concentrations alongside 'process efficiency' (e.g. low-carbon energy generation) and 'consumption curtailment' (e.g. energy efficiency). However, the most widely known term relating to carbon capture is 'carbon offsetting'. This involves estimating CO_2 emissions associated with a given action (e.g. running a car for a year) and 'neutralising' this by funding positive activities such as tree planting or renewable energy projects. In 2006, companies and individuals in the UK spent around £4 million offsetting carbon emissions.[35]

Tree planting has encountered criticism for not being permanent carbon capture as a tree will release all the CO_2 it has absorbed from the atmosphere at the end of its life when it decays or is burnt. However, this is perhaps not taking the full picture into account as any new forest areas created are naturally self-sustaining. If replanting takes into account local water resources and uses diverse native species it can certainly be regarded as positive action yielding benefits with respect to biodiversity, flood protection, aesthetics, air quality, microclimates etc.

The concept of carbon offsetting has been criticized as it allows a 'business as usual' approach rather than encouraging systems to evolve. Clearly excusing behaviour which should be avoided in the first place is not ideal. However, offsetting remains a positive step individuals can take to move to a more sustainable way of living especially if combined with change.

CCS more specifically refers to an extensive variety of techniques that can be used to mitigate large-scale emissions where they are generated (e.g. direct treatment of flue gases from gas-fired power plants). Storage of the captured CO_2 has been proposed in geological features, such as unminable coal beds or depleted oil/gas reservoirs. Other lower-cost alternatives such as ocean storage are less permanent and can be potentially damaging to local ecosystems e.g. by increasing sea acidity. These treatments come at a significant cost. The IPCC reports that the best case they have studied for a mineral carbonization technique would require a 30–50 per cent increase in energy prices.[36] However, the potential is substantial and the IPCC has also estimated that CCS could contribute 15–55 per cent to the cumulative mitigation effort worldwide until 2100, although this technology has been slow to be adopted on any significant scale. One obstacle to the development of CCS may be the inability of this sort of technology to capture public imagination compared to say tree planting in which an individual can actively participate and also directly perceive a positive change.

Iron fertilization is one of the more novel carbon capture techniques. It involves 'seeding' areas of the ocean which have a low iron concentration with large quantities of iron particles. Iron in the upper layer of the sea, which receives direct sunlight, triggers photosynthesis through plankton growth. This not only absorbs CO_2 but stimulates local ecosystems.

Carbon offsetting is not an exact science and different companies calculate 'carbon footprints' using different methods. However, the salient point is that action is being funded to compensate for emissions. In the same way as 'renewable energy should pick up from where energy efficiency has finished', it can be said that 'carbon offsetting should pick up from where carbon-reducing practices have left off'. Although developing rapidly, this area is still in its infancy and the development of systems provided by regulatory bodies may help if they are not over-prescriptive.

Can we afford renewable energy and sustainability when economies are slow?

Sustainable development, which includes the use of renewable energy, can be viewed as luxury that often cannot be afforded even during the most economically buoyant periods. During economic downturns it may seem that ideals of sustainable development move further out of reach. However, the opposite may be true.

One of the main elements of sustainability aligns itself directly with the 'tightening of belts' required during economic recessions: improving resource efficiency. This includes reducing energy consumption (including on-site energy generation), lowering water consumption and waste generation, and a drive for more efficient procedures such as reducing the waste during construction. A tightening economy could, therefore, see sustainability move from the fringes into standard design and develop thinking to take advantage of streamlined practices and economies of scale. Steady escalation of energy prices and increasing regulation (e.g. building performance certification) should also help drive this change.

Similarly, the energy industries may follow suit in order to decrease costs and increase profitability as the prices they have to pay for raw resources increase. The current direction power companies take follows the 'high leverage' commercial mentality where prices are kept competitive and high volumes of sales are encouraged, i.e. 'sell as much as possible to make as much profit'. This approach is understandable when raw resources are cheap and plentiful. However, a reversal into the other commercial mentality may occur whereby the price per unit increases and low volumes are sold. This could be beneficial for all concerned. For example, reversing pricing schemes so energy costs increase with customer use will encourage customers to think seriously about energy use and to take active measures to reduce consumption. Price increases will also help drive the move to invest in local renewable energy and to set up efficient decentralized energy networks. Reducing generation costs (operation cost and resource consumption) will also help energy companies to move through the up-and-coming transition period where fossil fuels are gradually phased out.

It should be kept in mind that standard economic comparisons of current centralized electricity generation technologies to efficient decentralized energy networks tend to omit wider environmental and societal impacts. The cost of impacts such as spillages, leaks, air pollution, acid rain and long-term health damage are difficult to account for with any degree of certainty. Another reason that these costs are not generally taken into account is that the developer does not have to pay for the damage associated with these secondary effects. In addition, any evaluation of these impacts will tend to be underestimated, as secondary knock-on effects are even more elusive to quantify.

SUMMARY

Safe, secure, sustainable and cost-effective long-term energy supply is required across the globe. While reducing energy consumption may be part of the solution (and there is considerable scope for this), issues relating to the most appropriate way forward for energy generation still remain. At this juncture, from the wider perspectives that have been outlined, some fundamental questions can be raised:

- Is it appropriate to remain dependent on oil/gas for electricity generation?
- To what extent is nuclear energy necessary and acceptable?
- Is a large-scale return to coal-fired electricity generation appropriate?
- Should resources be spent on carbon sequestration (of fossil fuel-fired power stations)?
- How should energy efficiency be handled?
- Should carbon offsetting programmes be carried out?
- Should an investment be made in a portfolio of renewable energy technologies?
- Should local decentralized energy networks be implemented?
- Should wind energy and urban wind energy development be supported?

Wind energy is by no means a straightforward option. However, urban wind energy can be a viable way to provide emission-free energy generation (if renewable energy is used to power the installation) although several issues may need to be reconciled. A number of these issues, which relate not only to economics but to subjective and ethical aspects, are summarized in Figure 1.7.

Economics is usually one of the first issues to be raised when discussing urban wind energy and this is a subject addressed again in this text. However, it may be useful to consider an alternative point of view: 'If we have the technological know-how to address a given problem, what role should economics play?' Is it economical to trade fairly; to ensure our food does not contain potentially harmful additives; to save endangered species; or to stop collecting compound interest from developing countries when the original loan has been paid back?

Economic viability for a simple large-scale stand-alone turbine has been crudely demonstrated. However, payback periods can vary hugely and there will be cases, especially where wind resources are poor, where this period will extend far beyond the design life of the turbine. Economics of wind turbines can be evaluated based on return on investment (ROI). Wind energy will fare poorly with this type of comparison unless the increase in cost of energy is well above inflation (e.g. >10 per cent). The ROI viewpoint is relatively short-sighted, as the model of investment purely for maximum personal gain is not in line with ideas of sustainability. As the future unfolds one may hope to see investments being made in ever-widening circles, considering individual wealth as part of community, regional, national and global wealth.

We may never be able to predict the outcome of environmental problems such as climate change with any high degree of certainty and this situation makes selecting a direction with any high degree of confidence difficult (especially when action consumes time and money). However, there may now be enough certainty to say: 'We think there are major issues with potential widespread negative impacts, so we need to devise and implement long-term solutions in all relevant areas.'

While considering these issues, abundant wind resources are available and offer substantial opportunities to at least partly address these key energy supply issues. The next section focuses on ways available to tap into this energy resource.

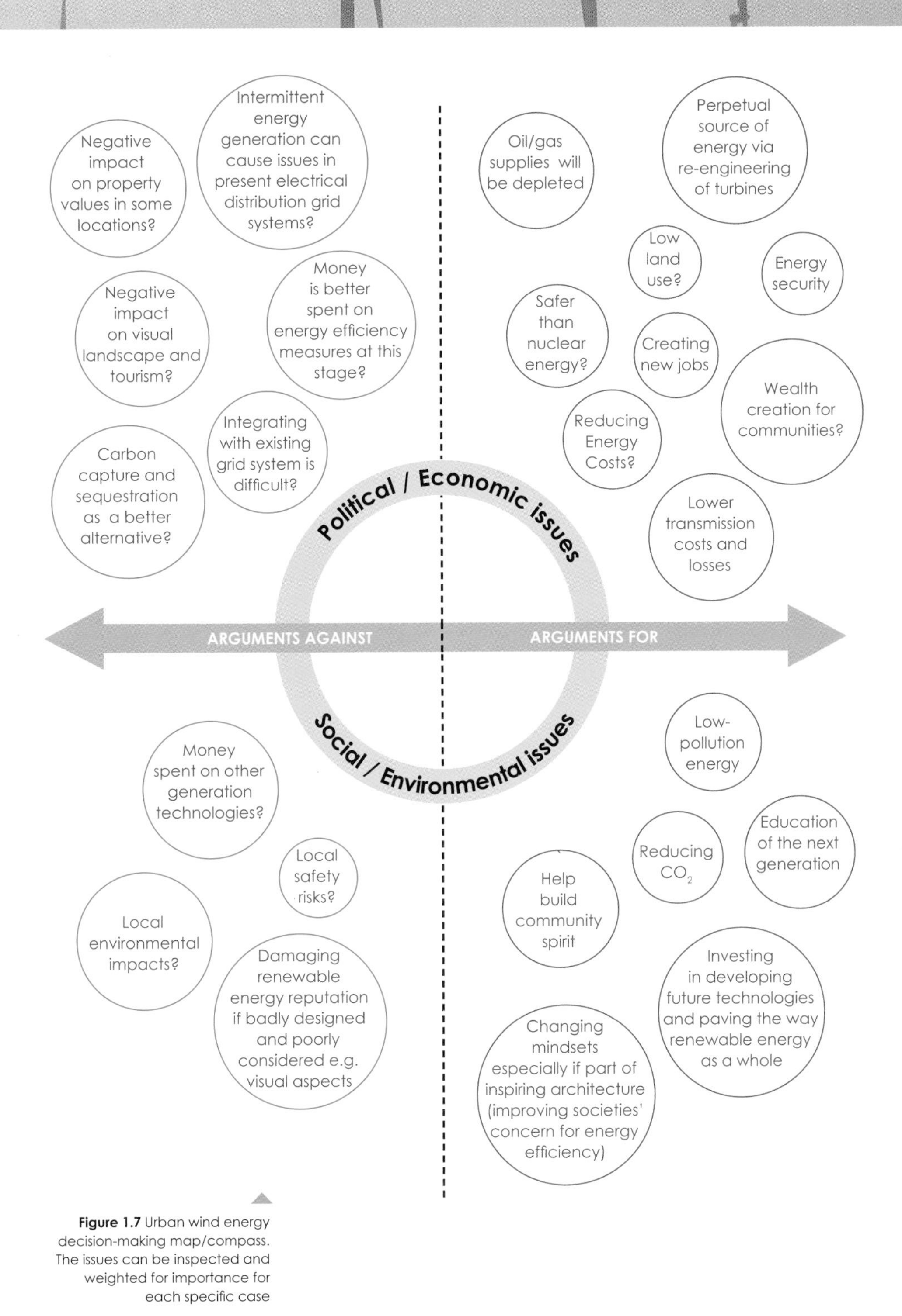

Figure 1.7 Urban wind energy decision-making map/compass. The issues can be inspected and weighted for importance for each specific case

REFERENCES

1 *IEA Key World Energy Statistics 2008*, International Energy Agency (IEA).

2 *BP Statistical Review of World Energy* (June 2009).

3 *IEA World Energy Outlook 2008*, International Energy Agency (IEA).

4 *GWEC – Global Wind 2008 Report*, Global Wind Energy Council (GWEC).

5 BP Statistical Review of World Energy (June 2009).

6 *GWEC – Global Wind 2008 Report*, Global Wind Energy Council (GWEC).

7 *GWEC – Global Wind 2008 Report*, Global Wind Energy Council (GWEC).

8 *GWEC – Global Wind 2008 Report*, Global Wind Energy Council (GWEC).

9 *GWEC – Global Wind 2008 Report*, Global Wind Energy Council (GWEC).

10 *AWEA Small Wind Turbine Global Market Study*, American Wind Energy Association (May 2009).

11 *AWEA Small Wind Turbine Global Market Study*, American Wind Energy Association (May 2009).

12 *BWEA Small Wind Systems – UK Market Report 2009*, British Wind Energy Association (April 2009).

13 *AWEA Small Wind Turbine Global Market Study*, American Wind Energy Association (May 2009).

14 *BWEA Small Wind Systems – UK Market Report 2009*, British Wind Energy Association (April 2009).

15 http://epp.eurostat.ec.europa.eu (July 2008)

16 http://tonto.eia.doe.gov/dnav/pet/hist/rbrteM.htm (June 2009)

17 *BP Statistical Review of World Energy 2006*

18 *UK Energy Policy: The Small Business Perspective and the Impact on the Rural Economy*, Small Business Council (February 2006)

19 Amory B. Lovins et al, *Small is Profitable: The Hidden Economic Benefits of Making Electrical Resources the Right Size*, Earthscan (2003)

20 http://politics.guardian.co.uk/interviews/story/0,,1516676,00.html#article_continue

21 Passiv Haus Institut, Rheinstr. 44/46, D-64283 Darmstadt, Germany (www.passiv.de/)

22 www.cepheus.de

23 *UK Energy Policy: The Small Business Perspective and the Impact on the Rural Economy*, The Small Business Council (February 2006)

24 *Canada's VRB Energy Storage System*, Tapbury Management of Donegal (Refocus Nov/Dec 06)

25 http://en.wikipedia.org/wiki/Intermittent_power_source

26 *Investigation into the Potential Impact of Wind Turbines on Tourism in Scotland*, VisitScotland (2002)

27 European Commission, *External Costs: Research Results on Socio-environmental Damages due to Electricity and Transport* http://ec.europa.eu/research/energy/pdf/externe_en.pdf (2003)

28 Nordhaus, W. D., and Boyer, J., *Warming the World: Economic Models of Global Warming*, MIT Press, Cambridge, MA (2000)

29 N. Caillon, J.P. Severinghaus, J. Jouzel, J.-M. Barnola, J. Kang, and V.Y. Lipenkov, Timing of atmospheric CO_2 and Antarctic temperature changes across Termination III, *Science*, 299(2003) pp1728–1731

30 *New Scientist*, 'Climate Myths' (19 May 2007) p42

31 *Can Nuclear Power Provide Energy for the Future: Would it Solve the CO_2-Emission Problem?* Jan Willem Storm van Leeuwen and Philip Smith, (June 2002)(www.greatchange.org)

32 British Wind Energy Association (www.bwea.co.uk)

33 Defra, *Review of the Climate Change Programme*, Consultation Paper, (December 2004) p42

34 *Wind Power in the UK: A Guide to the Key Issues Surrounding Onshore Wind Power Development in the UK*, Sustainable Development Commission (www.sd-commission.org.uk) (2005) p18

35 www.guardian.co.uk/science/story/0,3605,1673265,00.html (23 December 2005)

36 *Carbon Dioxide Capture and Storage*, IPCC (Intergovernmental Panel on Climate Change) (www.ipcc.ch/activity/srccs/SRCCS.pdf) (2005) p321

Urban Wind Energy Potential

Following the success of nearby Hull 1 (660kW turbine installed in 2001) the Hull 2 turbine was installed in Hull, Massachsetts in May 2006 (a Vestas V80 1.8MW producing 4,500MWh/a with a total cost of US$3 million) and has a 95 per cent residents approval. (Andrew Stern & Malcolm Brown, hullwind.org)

2

INTRODUCTION

WHERE THERE'S A WIND THERE'S A WAY: DESIGN OPTIONS AND OPPORTUNITIES

The scope for integrating wind energy in urban areas with good wind resources is extensive. Three main categories of project can be identified: small wind and retrofitting, large-scale stand-alone turbines and building-integrated turbines (where the buildings are shaped with wind energy in mind).

This section explores the path being created by pioneers and early adopters and presents a number of examples for each category.

Classifying urban wind energy into distinct categories can serve to aid the decision-making process by narrowing down the amount of specialist technical material that needs to be absorbed. For example, an owner of a building in a city would be able to focus on the relevant area of retrofitting and leave aside the others.

Key concerns to be addressed at the feasibility stage include energy yields (including percentage of annual building energy demand offset), environmental impacts and first costs/return on investment.

There are several other schemes that fall outside these main categories and are useful in removing any preconceived ideas of what urban wind energy should look like. These are presented at the end of this section, where attention is turned to future and emerging trends. This includes a review of the latest technological innovations being developed and used by designers.

Therefore, this section covers:

1) Small wind energy: 'Retrofitting' and building-mounted wind turbines
2) Large wind energy: Stand-alone wind turbines
3) Building-integrated wind turbines
4) The future of urban wind energy

To aid fuller understanding of urban wind energy potential, some fundamental principles explaining how wind turbines work are first summarized in Box 2.1.

BOX 2.1

THE POTENTIAL EXTRACTABLE POWER CONTENT OF THE WIND

The power in 'free-flowing' wind (i.e. not locally accelerated) is given by the well-known kinetic power term $\frac{1}{2}\dot{m}.v^2$ where $\dot{m}$ is the mass flow rate (kg/s) of the air passing through the swept area of the turbine blades and v is the velocity of the free wind (m/s). For convenience the wind turbine power equation is expressed in terms of swept area. Therefore the mass m is replaced with ρAv where ρ is the density of the air (kg/m^3) and A is the swept area of the blades (m^2).

Wind turbine power equation:

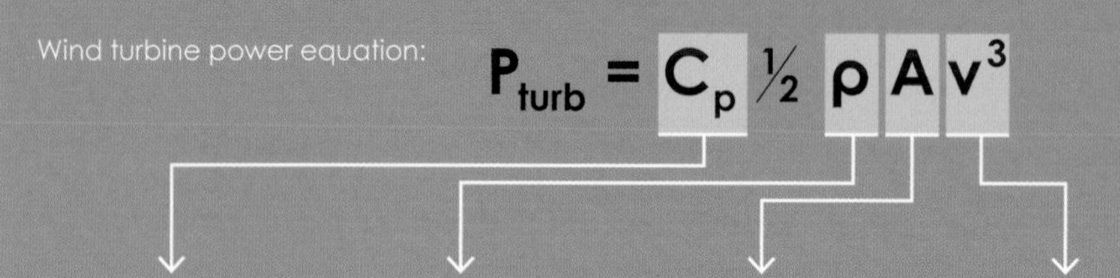

$$P_{turb} = C_p \tfrac{1}{2} \rho A v^3$$

The coefficient of performance C_p
This depends on the specific turbine. The maximum theoretically value is called the Betz limit and is 0.59. This coefficient varies for different wind speeds. A typical value is around 0.3. It can be as high as 0.5 for well-designed blades, as shown in Part 4, although for 'drag type' turbines this can be around 0.1 for higher wind speeds.

Air density
This can be an important variable to consider if a turbine is located in cold or warm climates. The air density decreases with increased temperature (and humidity) and increases with decreased temperature. The height of the turbine above sea level can also be considered in some cases as the air gets thinner at high altitudes.

Swept area of the blades
This is also an important variable as doubling the swept area A doubles the energy output.

Free wind velocity
A very important variable as the power P is a cubic function of v – i.e. doubling the wind speed multiplies the energy by a factor of 8.

For example: For a wind speed of 6m/s, a swept area of 3850m² (blade diameter of 70m), an air density of 1.2kg/m³ (air temperature of 20°C at sea level) the power in the wind is:

$$P_{wind} = \tfrac{1}{2} \times 1.2 \times 3850 \times 6^3 \approx \mathbf{500{,}000}\,\mathrm{W}$$

If the coefficient of performance is taken to be 0.3 the turbine can extract 30 per cent of this energy in the wind:

$$P_{turb} = 0.3 \times 498{,}960\,\mathrm{W} \approx \mathbf{150}\,\mathrm{kW}$$

To relate these figures to more concrete examples, this is enough power (instantaneous rate of energy production) to run either:

- 50,000 A-rated fridge-freezers (3W average)
- 30,000 mobile phone chargers (5W)
- 7500 energy-efficient light bulbs (20W)
- 5000 computers (30W)
- 1500 TVs/stereos (100W)
- 190 microwaves (800W)
- 125 vacuum cleaners (1.5kW) or
- 60 kettles (2.5kW)

In the latter part of the 20th century, energy consumption was not high on the list of concerns for manufacturers of domestic goods. However, these manufacturers are now tuning into energy efficiency as evidenced by the introduction of A-rated energy-efficiency appliances. In this context, local renewable energy production is becoming more relevant and able to deliver a greater percentage of local energy needs.

It should be noted that the wind power equation is also the basis for the energy production calculation which is most commonly evaluated on an annual averaged basis (as shown in Table 1.2). Multiplying the power (kW), taken at the mean annual wind speed for a given site, by the number of hours in a year (8760h/y) will give the mean annual energy (kWh/y). However, it is important to take into account the wind speed distribution without which the annual energy output calculaution can be significantly underestimated. This is explained further in Part 3. Other factors such as the availability of the turbine and the variation of the coefficient of performance with wind speed should also be taken into account (these are also addressed in subsequent sections).

1) SMALL WIND ENERGY: RETROFITTING AND BUILDING-MOUNTED WIND TURBINES

The words 'urban wind energy' may typically conjure up pictures of very small, three-bladed horizontal axis wind turbines (HAWTs) on short poles attached to houses (Figure 2.1). Paradoxically this image may be the least realistic.

Currently, home-mounted wind turbines are perhaps best suited to enthusiasts as they require a relatively high level of knowledge and investment in time in order to avoid unnecessary complications and poor energy yields. This includes understanding key issues such as predicting available wind resources, avoiding turbulence, mitigating environmental impacts, preventing structural damage and understanding the economic aspects. It will also often involve applying for grants, obtaining planning permission, dealing with the production of the electricity (e.g. selling to the grid) as well as maintenance issues.

Figure 2.1 Small home-mounted turbine in Scotland from UK manufacturer Windsave, rated at 1kW and with a 1.9m blade diameter (Stuwart Russell)

Figure 2.2 Swift (UK) turbines rated at 1.5kW with a 2m blade diameter, mounted on a Sigma Home (Renewable Devices)

There are a number of concerns frequently expressed in relation to home-mounted wind turbines. Most relate to manufacturers (and planners) encouraging/allowing homeowners, who do not fully understand all of the key issues, to erect small turbines on their roofs (which can lead to disappointment and frustration). Generally, standard walls and chimneys are not built to withstand the types of prolonged stresses originating from a turbine mast. Therefore, these structures, unless specifically designed for the purpose, or assessed by structural experts, should not be used as they could present a real safety threat, e.g. from falling bricks, tiles or turbines.

The potential for the misuse of 'small wind' to do unnecessary damage to the reputation of urban wind energy as a whole is therefore real and substantial. This harm could extend to non-urban wind energy or renewable energies in general.

One of the major issues relates to the quality of wind resources at rooftop level. Not only are these much slower than in 'open-field' sites but they are less uniform. The swirling turbulent character of low-level urban wind, due to the complex interaction of the wind with buildings, is detrimental to the amount of energy that can be extracted (as expanded on in Parts 3 and 4). These swirling flows also exert irregular stresses on the blades and so also reduce the longevity of the turbine (although the extent of this is difficult to quantify as different turbines respond differently).

Figure 2.3 Small stand-alone turbine at the Mile End Ecology Centre, London, from UK manufacturer Proven, rated at 6kW with a 5.5m blade diameter (Ioannis Rizos)

Despite the lack of widespread applicability there is still some potential for home-mounted wind turbines and the subject is still worthy of discussion. In ideal conditions, where the turbine can access winds proved to have an annual mean wind speed of 5.5m/s, a 2m diameter HAWT can generate about 40 per cent of the electric energy demand from a typical UK three-bed home or 40 per cent of the total energy demands of a 100m^2 house meeting the Passiv Haus standards (as presented in Table 1.2).

Generally significant improvements can be gained from having taller supporting poles. Images which come to mind when considering future environmentally friendly homes with wind turbines should picture the turbine very high above the roof level. Taller masts will of course have cost implications and create additional environmental issues such as increased visual impacts which some planners may contest.

A growing number of small manufacturers are producing wind turbines in an attempt to break into this potentially large market. The most common are horizontal axis wind turbines. Although electricity 'microgeneration' can be thought of as electricity production below 50kW, the most common HAWTs are typically rated at around 1kW (with a blade diameter around 2m – see Figure 2.2) or around 5kW (with a blade diameter around 5m – see Figure 2.3). Vertical axis wind turbines (VAWTs) are also available, usually up to 5kW. Appendix 1 provides a list of potential manufacturers.

Although there are now many summaries available on basic points relating to small wind turbines, further consolidation is required. Several groups have made efforts towards meeting this need and to encourage the development of this new sector.[1,2] The British Wind Energy Association (BWEA), for example, are developing standards and carrying out testing in order to allow easier comparisons of available technology (standardizing rated wind speed and annual yield criteria). Although useful, this work applies to ideal conditions. It will not address how these turbines perform in reality where winds may not consistently meet the blades at the ideal angle, i.e. during periods of high wind turbulence. Comprehensive texts are also available on the practicalities associated with the use of small wind, e.g. Gipe[3], who reiterates the common concerns about the poor wind energy resources in many urban areas especially for turbines mounted on houses on short poles.

Even though there are now a growing number of examples of urban wind turbines, to date, a full picture of the level of success of these endeavours is only just beginning to emerge. This is largely due to the long timescales involved when monitoring energy outputs and average site wind speeds.

Beama Energy has compiled performance data for nine small HAWTs as part of a microgeneration assessment project. The data show a large range of 'capacity factors': between 1 and 15 per cent (with an average of 8 per cent). Those performing above average still have a capacity factor half those of large turbines in wind farms operating in open fields. This may not be surprising as the coefficient of performance of some small-scale turbines will be lower than those of large-scale turbines and the hub heights of the majority of these smaller turbines are low compared to hub heights of around 70m for large-scale turbines. No 'expected capacity factor' is given, which would provide information on the quality of the wind resource, and so it is difficult to say whether low performance is mostly due to the design of the turbine or the positioning and quality of the local wind resource.[4]

The Warwick Wind Trials[5] have also provided results from small building-mounted HAWTs. The performance of the majority of the 30 sites has been poor, with the main reason being attributed to the low wind speed at the level/position of the turbine. A notable discrepancy between manufacturers' performance curves ('power curves') and the measured performance curves has also been identified. The averaged energy being produced per day has been around 200Wh with the energy consumption (e.g. to power the inverter) being over 100Wh per day in some cases but averaging around 60Wh per day. Therefore, with some home-mounted wind turbine installations located in unsuitable areas more energy has been consumed than generated. The study reiterates the need for wind resource prospecting (as discussed in Part 3).

Data on the performance of VAWTs in urban areas are currently not widely available although this technology could prove a significant contender to the historically favoured and cheaper HAWTs. VAWTs can be found in a numerous forms and have several advantages. For example, they can deal with changes in wind direction more easily, they can be perceived as more aesthetically pleasing and less audible (as the 'tip speed ratio' is low). Furthermore, although more material is required per m^2

of swept area than with a HAWT, there is still significant scope for the continuation of the downward trend in equipment costs if uptake can increase. However, there are other issues to consider e.g. 'overspeed' control for Darrieus type VAWTs (i.e. how to avoid damage to the turbine in wind speeds >12m/s).

Generally, the advice to homeowners wishing to make use of wind energy would be to exercise a degree of caution – if in doubt, do not proceed (for example, if the quality of the available local wind resource is not known). Of course, it is natural for a newcomer to any subject to underestimate the associated complexities and so the onus is on local planning departments to provide simple guidance at the very early stages or on their websites. A simple series of questions can quickly give someone an idea of levels of knowledge of the general principles involved. For example:

- What is the coefficient of performance of a turbine and why is it important?
- What are the methods of estimating local wind resources and what is a Rayleigh distribution?
- What is turbulence, why is it important and where are high-turbulence zones found?

There has been a mixed response to the availability of 'off-the-shelf' wind turbines such as those available from a popular DIY chain store in the UK. Although a pre-survey has been offered and accredited installers have been used, there is still scope for inappropriate use. Restricting the sale of these devices has been suggested and this is often coupled with the idea that planning controls should in fact be tightened in certain areas, and not made easier, to stop those without suitable expertise potentially causing environmental damage and harming the image of wind energy. One positive aspect of this activity, however, is that bringing this technology direct to the public has generated considerable interest. This in turn serves to excite the market, encouraging investigation and investment. In a climate of increasing demand for renewable energies this could lead to:

- improved turbine technology being developed;
- lower costs;
- better performance data;
- more comprehensive standards being developed;
- better guidance for planning and grid connection;
- a database of wind resources emerging.

The price to pay for this advancement may be a number of dissatisfied customers. Of course, if dissatisfaction reaches a certain level then the subject of home-mounted turbines could retreat into hibernation.

As wind turbines installed at high levels are able to access winds with higher energy contents, this technology naturally lends itself to use on taller buildings. One example is the London Climate Change Agency (LCCA) set up by the Mayor of London in June 2005 to tackle climate change through promoting renewable and sustainable energy, which is housed in three floors of Will Alsop's 'Palestra' building, Blackfriars Road, London. The attempt to 'show by doing' has seen fourteen 2m diameter Swift turbines installed in 2006 as part of an integrated solar/wind system (63kWp PV array) at a total cost in excess of £400,000. It will reportedly generate 3,397,000kWh and reduce CO_2 emissions by 3300 tonnes during its lifetime. However, teething troubles have seen the turbines removed only months after normal operation began when the manufacturer issued a call-back due to component failure.

Figure 2.4 WT6000 5.5m wind turbine at a BP station, Wandsworth, London, by UK manufacturer Proven (Ivan Jovanovic)

Although turbines on taller buildings can reach better quality winds, the disparity between energy demand and energy production, as seen on a per building basis, increases significantly. However, larger turbines can be used, such as the two Proven 6kW turbines installed on the roof of the Innovations Centre in Plymouth, UK. Proven is a well-established UK small turbine manufacturer and has several hundreds of installations in the UK alone, many sited in the built environment (as shown in Figure 2.4). They are 'downwind' turbines – that is, the blades are 'downwind' of the nacelle (most turbines have their blades facing into the wind). The blades are flexible and fold during periods of very high wind speeds to prevent damage to the turbine. The two largest Proven machines are the WT6000 (5.5m diameter blades, 6kW) and the WT15000 (9.4m blade diameter, 15kW) turbines with tower heights of 9m for the WT6000 and 15m or 25m for the WT15000. The maintenance required for these types of turbines are only around four hours per year, in order to grease the bearings.

A Proven 6kW turbine is also installed on the site of the Mile End Ecology Centre, London, shown in Figure 2.3. The main Eco-centre building is approximately 40m from this device and the row of houses adjacent to the park are at a distance of 90m from the turbine. The hub height has been limited to 9m and this of course highlights one of the main issues of siting urban turbines: access to good wind resources. Should planning conditions result in a seemingly over-restrictive limit on the allowable hub height, they can be countered by submitting anemometer data recorded from two heights to highlight the decrease in the time of the year the turbines are turning and the loss in energy yields.

As shown in Figure 2.4, large buildings can place wind turbines in 'wind shadows' robbing them of energy from certain wind directions. Projects should be well considered in order, for example, to avoid installations becoming merely 'aesthetic gimmicks' if wind resources in a certain area are poor, or claims of 'greenwashing' where, in some cases, companies may be perceived as trying to detract from their core activities (which may be non-sustainable) with highly visible green technologies (regardless of their actual intention).

Perhaps one of the most well-known introductions to urban wind energy was during the World Expo held in Hanover Messe, Germany in 2000, where the Dutch presented several Tulipo mounted on their six-storey pavilion designed with MVDRV architects. Since then, they have installed several turbines in the built environment, e.g. at Knivsta High School in Sweden (see Figure 2.5).

Figure 2.5 Tulipo turbine at Knivsta High School in Sweden with a 5m blade diameter, rated at 2.5kW
(WES – Wind Energy Solutions)

The Tulipo has been designed with aesthetics in mind (Figure 2.6) and WES presents it as a 'certified urban turbine' as it holds an American safety certificate. A Tulipo was installed at the Blackburn Enterprise Centre, UK in November 2007 and the location chosen for high visibility to promote wind energy within the community. The turbine is connected 'behind the meter' to decrease the electricity bill of the centre. It has an estimated 15-year design life, and although it is not the most efficient design (rated at 2.5kW) and has an active yaw mechanism (requiring electricity to rotate the blades into the wind, which is uncommon for small turbines), it is reported to produce 10,000kWh/a on sites with 6–6.5m/s at hub height. It emits only 35dBA at 20m from the source during winds of 9m/s ,when the blades rotate at 140rpm, and 72dBA at the nacelle (Part 3 contains information on acoustics). The standard tower heights are 12.25m or 6.25m.

Figure 2.6 WES5 Tulipo turbine with access ladder
(WES – Wind Energy Solutions)

As mentioned, the disparity between manufacturer energy yield predictions based on a specific mean wind speed and the actual energy output as a result of the actual available wind resource (which can be significantly lower) is one of the main sources of concerns for small wind – as pre-installation on-site wind monitoring is often not carried out. However, wind monitoring may not be as daunting a prospect for homeowners as first considered (as discussed in Part 3) and even if wind resources are lower than recommended speeds a turbine can still produce useful amounts of energy.

An assessment of small wind turbine performance has been carried out by the progressive team at the Hockerton Housing Project, UK.[6] Their turbines, a 5.5m diameter Proven rated at 6kW and a 5.4m diameter Iskra rated at 5kW, produce around 4.6MWh/a and 4MWh/a respectively. Both manufacturers claim their turbines will generate around 12MWh on sites with annual mean wind speeds of 6m/s at hub height. However, the Hockerton anemometer, at 18m above ground level, has revealed the annual mean wind speeds to be around 3m/s. The turbines are on guyed towers 26m above ground level and the extra height allows the turbines to access wind speeds approximately 0.5–1m/s higher.

Interestingly, the Hockerton team has reported that the planning process for first turbine, commissioned in 2002, took around five years (with four applications and three appeals) after concerns of visual intrusion, noise disturbance to the local residences and distraction to drivers on the road some 250m from the turbine. The planning application for the second turbine, commissioned in 2005 after the first had been operating for over two years, took 11 weeks as very few objections to the application were put forward.

Despite the low winds, the turbines are expected to produce 80 times their embodied energy, and 4MWh/y is the equivalent of running around 140 compact fluorescent light bulbs (20W) for four hours a day every day. Payback for this site is around 15 years (Hockerton were awarded a 40 per cent capital grant); however, the team considers their renewables as part of the houses, not as a business venture.

An interesting example of an urban turbine is the Skystream installed on La Case Verde in California. The turbine has a blade diameter of 3.7m and is rated at 1.8kW. In this windy location it is reported to produce around 600kWh/month.

Figure 2.7 Skystream 3.7 from US manufacturer Southwest Windpower as part of La Case Verde, a sustainable residence made from recycled materials in San Francisco, California. The full installation, including permits, was around US$20,000. (Robin Wilson)

2) LARGE WIND ENERGY: STAND-ALONE WIND TURBINES

Decades of experience and investment in large stand-alone wind turbine technology (blade diameter > 20m) has culminated in a well-accepted form and mode of operation: three 'pitch control' blades on a horizontal axis. The potential for green on-site energy production offered by these multi-megawatt turbines used in wind farms (Figure 2.8) in windy regions is now considerable. Turbine reliability, performance and value are continually improving. This technology has been at the stage for a number of years where it is more than viable (in certain locations). As mentioned, this technology can become 'carbon positive' well within the first year of operation and payback times can be as short as three years, after which a steady source of income will be enjoyed. Long-term experience in large-scale turbines in urban areas is still currently lacking (traditional wind farm developers are attracted to developments which install several tens of turbines).

Although wind speeds are generally lower in built-up areas, large-scale urban wind energy can be successfully implemented (as shown by the examples in this section) if site wind monitoring reveals adequate resources and environmental impacts are demonstrably low. These can be found in many urban areas especially in elevated or coastal locations.

Figure 2.8 Large-scale multi-megawatt turbines (Gamesa)

'To renewable-energy supporters, the wind turbine symbolizes the hope of a green, clean future, but to opponents, they might as well be Martian tripods from War of the Worlds, advancing inexorably across our precious countryside' (Steve Rose).[7]

Moving large-scale wind turbines into the built environment has several advantages over traditional remote wind farms:

- transmission losses are minimized (which helps to address the loss in wind resources between urban and open sites);
- transmission tower costs are removed and cabling costs are reduced;
- access road costs are also reduced or eliminated;
- income is improved as electricity can be sold directly to the end users (instead of feeding to the grid);
- green/wildlife areas can remain 'unspoiled';
- there can be a lower impact on biodiversity.

The disadvantages mainly relate to potential local environmental impacts which may occur in certain sites (as described in Part 3). However, these can be suitably managed in many (but not all) cases.

From the number of successful installations of large-scale wind turbines seen in urban areas over recent years it is becoming increasingly clear that the potential for significant wind energy generation on our doorsteps (where the energy is required and consumed) could be high.

Figure 2.9 Green Park 2MW Turbine, Reading, UK, with a 70m diameter, by German manufacturer Enercon (Ecotricity)

Generally these larger turbines can be sited very close to buildings and roads. One example is the Green Park turbine located on South Oak Way, Reading, UK, erected in 2005. Green Park is an expanding development of high-quality business and retail units, and the turbine has a 70m blade diameter, a hub height of 85m and is rated at 2MW. The building directly adjacent to the turbine (occupied by Cisco Systems) is only 70m away from the turbine. It is also very close (less than 150m) to a major motorway (M4, near junction 11), as shown in Figure 2.9.

Another example can be found in the Leonardo Da Vinci School in Calais, France. This turbine, installed in 2001, has a blade diameter of 20m, a hub height of 35m, is rated 130kW and located about 20m from the nearest school building (kitchens), as can be seen in Figure 2.10. Classrooms are only few dozen metres away.

The project cost was €160,000. The predicted average annual wind speed was around 6m/s and the estimated energy yield was 150,000kWh/a. The actual recorded annual energy produced has been 155,990kWh (with the energy being used for the school). This is a reasonable achievement as, although the school is only a few kilometres from the sea, the area surrounding the site for 1–3km would be characterized as 'sub-urban terrain', i.e. relatively high 'aerodynamic roughness' (these terms are expanded on in Part 3).

Both of these cases have the turbine about one blade diameter from the nearest building. However, it should be noted that these buildings are not residential buildings. In all cases the distance of turbines from residential building should be strictly limited. Although impacts such as noise or blade flicker can be easily tolerated for short periods of time, local homeowners, who have no means of relief, can become more sensitized. This is more likely to be the case if, in addition, they can perceive no tangible benefits from the turbine.

Although general rules of thumb are difficult to apply to wind energy, at this stage it may be useful to suggest one could begin by considering that residential areas should be separated from wind turbines by at least seven blade diameters. The Leonardo Da Vinci School has reported there have been no complaints related to the turbine or parents expressing concerns. The closest residential areas are about 150m away from the turbine (i.e. around seven blade diameters).

Figure 2.10 Leonardo Da Vinci School in Calais, France with a 20m diameter (hub height 35m) turbine rated at 130kW, from Dutch manufacturer Vestas (Frédéric Allard)

When any new large-scale urban wind turbine is proposed some reservations concerning potential problems that could arise, e.g. relating to visual impacts, safety and devaluing of properties (as discussed in Part 3), will be present. However, there are now more and more large-scale turbines appearing near buildings. As experience with the integration of this technology increases, information on what type of turbines, locations and mitigation methods are most appropriate also become clearer. Public response and feedback can also be gauged. In parallel to the increase in experience, awareness of 'the environment' is also increasing. This may mean more people will want to see this technology present in their daily lives.

A free-standing urban wind turbine can be a relatively simple option as it can be procured in an 'off-the-shelf' manner suitable for developers, investors, ESCos (energy service companies) and community schemes. If the impacts of installing wind energy are demonstrably low and local wind resources demonstrably high, a stand-alone wind turbine offers a very real means of addressing concerns such as energy security, pollution, and of course climate change while producing an attractive source of income.

Integrating wind energy into the built environment, if due care is taken in the design phase, cannot only provide a potential means to generate clean energy but serve as a constant reminder of the actions needed to mitigate environmental damage. An iconic wind energy generator, if implemented correctly, can also symbolize the 'spirit of humanity' to invest efforts in ensuring a positive future for the next generations.

Some more examples of large-scale stand-alone turbines are summarized below.

Figure 2.11 1.8MW Enercon turbine at the Ford Estate, Dagenham. London (Ecotricity)

Ford Estate in Dagenham, UK

April 2004 saw the construction of two 1.8MW wind turbines at the Ford Estate in Dagenham, with hub heights of up to 85m and 70m diameter blades (Figure 2.11). These power Ford's new Diesel Design Centre and were predicted to generate around 6.7 million kWh of electricity per year, which was published to be roughly the equivalent of the electricity consumption of 2000 homes.

One of the planning conditions (or 'Section 106' agreements) related to the local impact of blade flicker. The turbine is to be programmed to stop operating during certain times of day if this shadow flicker was perceived to be a problem for nearby residents.

The noise emissions report submitted as part of the planning application confirmed that the maximum noise level expected to be generated by the turbines is 38dB.[8] This is below the level of the existing noise environment which has the A13 road generating higher noise levels. The turbines are direct-drive (gearless), which limits the tones associated with mechanical sound.

Michelin Tyre Factory, Dundee, UK

In May 2006, two 71m diameter turbines were installed at the Michelin Tyre Factory, Dundee (Figure 2.12). The hub height is 85m and these two 2MW turbines are reported to generate enough electricity to power 2242 homes.

Figure 2.12 2MW Enercon turbine at the Michelin Tyre Factory, Dundee, UK (Ecotricity)

Sainsbury's Distribution Centre, East Kilbride, UK

In 2001, a turbine with a blade diameter of 44m was installed at the Sainsbury's Distribution Centre, East Kilbride, UK. The hub height is 65m and this 600kW turbine is reported to generate enough electricity to power 576 homes. The turbine is the first of Ecotricity's Merchant Wind Power (MWP) initiatives. Built without subsidy and supplied without a premium, MWP is a pioneering new concept whereby customers buy the electricity generated from their own on- or off-site wind turbine which is built, owned and operated by Ecotricity. MWP provides an economic and commercially viable source of renewable power for organizations with an environmental agenda.

Wood Green Animal Shelters, Cambridgeshire, UK

In 1990, a 26m diameter turbine was installed at the Wood Green Animal Shelter headquarters, on London Road, Godmanchester in Cambridgeshire (Figure 2.13). The hub height is 30m and it can be seen from 10 miles away.The cost of installing was £175,000 and it reportedly paid for itself in three years by selling the electricity to the local energy supplier. The turbine has a life expectancy of 25 years. Notably, this turbine is only 20m away from the nearest building (the local animal surgery).

Figure 2.13 26m diameter turbine at the Wood Green Animal Shelter, Cambridgeshire, UK (Wood Green Animal Shelter)

Antrim Area Hospital Wind Turbine, Northern Ireland

In 2005, a Vestas V47 turbine was installed at the Antrim Area Hospital, Northern Ireland. This 660kW turbine is reported to have the potential to provide enough electricity for the hospital during the night, and two-thirds of the power needed during the day, which would otherwise cost £90,000 per year. The turbine project cost £497,000. Conception to installation took three years. The civil works started in autumn 2004 and the turbine was delivered in January 2005 with only three days to install the turbine.

Great Lakes Science Center, Cleveland, USA

In 2006, the Great Lakes Science Center in Cleveland USA, installed a stand-alone turbine together with a photovoltaic array. The turbine has a blade diameter of 27m, is rated at 225kW and weighs 26 tonnes. It is expected to produce 7 per cent of the centre's annual electric needs. The turbine is 70m from the museum (which houses the real-time wind data display), less than 150m from the Cleveland Browns American football stadium and about 20m from the nearest road.

This turbine is re-engineered – i.e. taken from a wind farm being upgraded to much larger turbines and then re-engineered for reuse. In this case the turbine originated from a wind farm in Denmark and was shipped to the US to be refurbished.

Figure 2.14 Great Lakes Science Center 27m diameter turbine in Cleveland USA, rated at 225kW, from Dutch manufacturer Vestas (V27) (Jim Kolmus)

3) BUILDING-INTEGRATED WIND TURBINES

Building-integrated turbines, where 'buildings are designed with wind energy in mind', are an option for consideration by developers tuned into the change surrounding sustainable living.

Attempting to design a 'sustainable building' can appear paradoxical as construction uses resources and generates waste as does the operation of any new building during its lifetime. However, current design and construction practices, established in an era where energy and environmental issues were a peripheral concern, still have plenty of scope for improvement and the goal of the sustainable buildings remains a worthy ideal to move towards.

Building-integrated turbines are of course limited to new developments in relatively windy areas and will have natural constraints in the size of turbines they can accommodate. The vision behind integrating a turbine into a building, in some cases, is perhaps less a practical solution to be widely adopted than an architectural and cultural statement. The value of the possible cultural benefits should not be underestimated as architecture simultaneously reflects and influences culture and cultural changes. Having these powerful dynamic symbols integrated directly into the heart of urban communities could help change mindsets and have positive knock-on effects in terms of environmental action (e.g. homeowners improving energy efficiency or engaging directly in renewable energy).

Despite their more limited applicability (relative to large-scale stand-alone turbines), it is thought that they can be viable and efforts have been made in this area.[9] In 2000, Project WEB[10] gave the first comprehensively designed and tested example of a conceptual building-integrated turbine. The pioneering design of a twin-tower building with three integrated 35m diameter, 250kW horizontal axis wind turbines has now become an iconic form representing this field. Figure 2.15–2.17 present this design together with the first serious attempt which has since emerged to emulate these ideas – the World Trade Centre, Bahrain (constructed in 2007–2008).[11] This building has three 29m horizontal axis turbines suspended between two 34-storey towers of prime office space. Both of these towers have been designed to catch and accelerate the prevailing winds. In the case of Bahrain's World Trade Centre it should be noted that these are mild coastal winds and the building form is not fully aerodynamically optimised.

Figure 2.15 Pioneering conceptual building-integrated turbines from Project WEB (EC JOR3-CT98-0270), 1998-2000 (BDSP Partnership, Imperial College of Science Technology and Medicine - Department of Aeronautics, Mecal Applied Mechanics BV, University of Stuttgart - Institut für Baukonstruktion und Entwerfen Lehrstuhl 2 (ibk2))

Figure 2.16 First large-scale building-integrated turbine project World Trade Centre in Bahrain, 2008 (Ahmed Hussain)

Figure 2.17 Close-up of the World Trade Centre in Bahrain (Ahmed Hussain)

There may be a number of concerns which come to mind when considering these large-scale building-integrated turbines. Some require very careful consideration while others can be settled with some brief deliberation. At this early stage, before looking at some more examples of this technology, apprehensions relating to the following are discussed:

- energy yields;
- cost/value of investment;
- low energy generation to building energy use ratio and potential to mislead the public.

Figure 2.18 Physical testing of a prototype 2-storey building with integrated wind turbine from Project WEB (BDSP Partnership, MECAL, Xkwadraat (NL), CRLC RAL)

Energy yields – a potential concern?

Some designs for horizontal axis building-integrated turbines, as in Figure 2.19, require the blades to be fixed facing one direction (unlike conventional HAWT which can yaw into changing wind directions). For those cases where the blades are fixed, it should be noted that, although very little energy will be gathered when the wind blows from some directions, a well-designed building will accelerate or concentrate the wind from certain key directions. If there are strong prevailing winds then this type of technology can be viable if the building is appropriately shaped and orientated. Detailed examinations expanding on energy yields are presented in Part 5. However, noting some of the conclusions from the physical testing carried out in Project WEB, a non-yawing building integrating turbine can (if appropriately designed):

- accelerate winds (power enhancement) from winds +/- 75° from the direction of the prevailing wind (if the axis of the turbine is orientated with the prevailing wind);
- generate some energy even when winds are blowing 90° from the prevailing wind direction;
- generate at least twice as much energy from the prevailing wind than a 'free-standing' equivalent turbine.

Project WEB involved a high degree of large-scale physical testing using a turbine on a tower at 4.5m (see Figure 2.18). Although the results quoted above only apply to this particular geometry they give an indication of what can be achieved. It is expected that a further optimized geometry could outperform this configuration although it should be noted that a poorly conceived/tested geometry will fare much worse.

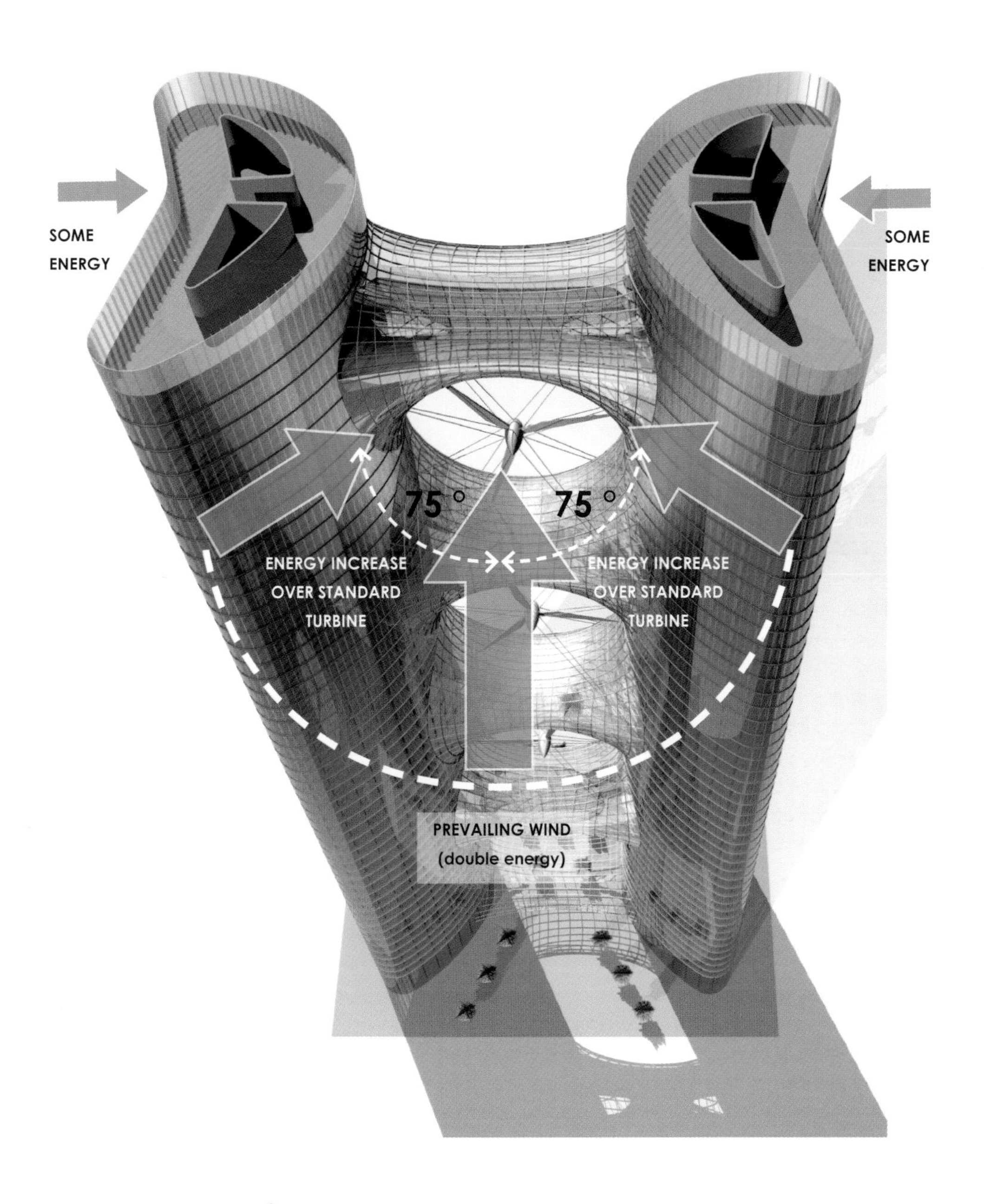

Figure 2.19 Energy yield increase for the Project WEB concept building from the results of large-scale physical testing and airflow models (ibk2 University of Stuttgart, BDSP Partnership)

As turbines do not start generating until the wind speed exceeds the 'cut-in' speed, any acceleration is welcomed. In urban areas, where winds are milder than open-field sites in the same region, accelerated winds can make all the difference.

It should be noted that shrouded or ducted turbines are generally considered in wind engineering circles to be poor performing technology and indeed stand-alone ducted turbines deliver only small improvements in energy yields for considerable additional costs. However, as buildings are large structures, building-integrated turbines can concentrate more airflow and produce more energy than stand-alone turbines (if, for example, the turbine is sized correctly within the shroud).

Value of investment – a potential concern?

Another main area of concern usually relates to the cost/value of integrating a turbine into a building. Besides the cost of the turbine, the additional structure and vibration control elements will increase costs as will any bespoke design or aerodynamic shaping of façades.

However, it is not all economic gloom and compared to the cost of conventional stand-alone turbines these additional costs will be offset to some degree by several factors:

- Fixed turbines are simpler and require less maintenance (as there is no 'yawing' mechanism).
- Tower costs can be reduced or eliminated.
- Foundation and access road costs are also eliminated.
- The cost of transmission towers and long-distance cabling are removed.

Concerns regarding a decrease in the value of the real estate which may be brought about by integrating turbines into a building are understandable, as it is conceivable that some areas nearer the blades could be potentially deemed unpleasant to occupy if poorly designed. Also there may be a tendency to think of the loss of potential 'floor area' by having a turbine and hence the loss in revenue. However, the real-estate value could equally increase as a result of the presence of turbines in a well-designed building. This is likely to be the case for buildings that are conceived to develop an 'iconic' status.

Figure 2.20 30 St Mary Axe, aka The 'Gherkin', London, which uses wind energy to drive the natural ventilation, modelled using CFD (streamlines coloured by velocity and façade coloured by pressures) (BDSP Partnership)

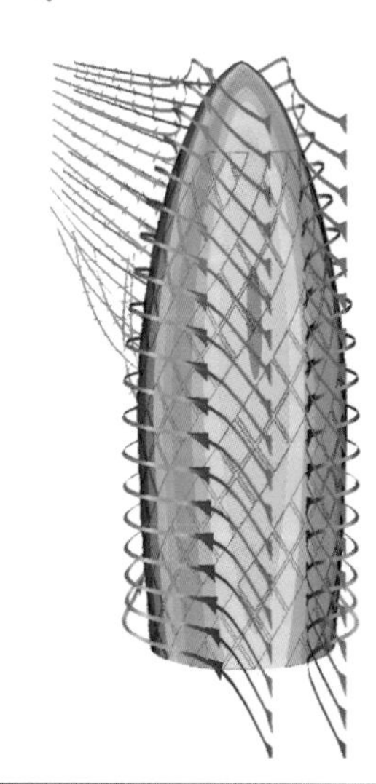

A simple example is 30 St Mary Axe, London, commonly referred to 'The Gherkin'. This building marries elegant iconic design with environmental functionality. This came at some significant expense over a 'standard' tower block but the investment and forward thinking from the design team has paid dividends both in capital and environmental terms. Built in 2003 it cost £138 million and was sold in 2006 for £600 million to German property firm IVG and UK investment firm Evans Randall. In this case wind energy was conceived to drive the natural ventilation through the three helical ventilation cores from which the unique façade design originates (these also provide natural daylight penetration into the deep core).

In addition to real-estate value benefits resulting from iconic status, the quality and feel of the space from a human point view is much higher than would be attained from cheaper characterless variations.

In general, the value of the floor space increases with perceived green-value of a building. Commercial companies who wish to position themselves as part of the emerging 'eco' orientated culture may be willing to pay a premium for genuine sustainable buildings. This also applies to potential homeowners who wish to be able to live in a home designed to facilitate more sustainable lifestyles. Indications of these premiums, where people are willing to pay more to contribute to sustainability, also extend into other areas such as organic or fair-trade produce, schemes linked to charities, ethical investment schemes and carbon off-setting.

Figure 2.21 The iconic 30 St Mary Axe, London

Changing mindsets of building users to take full advantage of sustainable features is another issue. Providing a well-designed building is necessary but not sufficient for the success of the design. Educating users to increase understanding and motivation to operate the building in the most effective way is also required. This idea relates to many areas and is true for wind energy. If generating the energy in the best way is the first part of the equation then the second part would be effective distribution and energy use.

Percentage of energy demand – a potential concern?
The amount of on-site energy generation is often expressed as a percentage of the total energy use (or total electrical energy use) of the building. For one small 2m diameter turbine supplying energy to one energy-efficient home the percentage can be high (as shown in Table 1.2). For a large stand-alone turbine, the amount of energy may be related to a new development or to a particular building such as a school, hospital, industrial plant or commercial area, in which case the percentage can again be significant. However, for building-integrated turbines, such as those located on top of high-rise buildings or integrated within or between large/tall buildings, the percentage of total energy use these devices will meet can be very low.

If only 1 or 2 per cent of the energy demand is being met, due to the type and large number of units tall buildings tend to contain, it may appear difficult to justify the expense. Furthermore, the percentage energy generation from the turbines integrated into a given building may be perceived by the occupants as high. This could lead to some occupants concluding that the 'energy issue' has been solved and it is fine to resort to a 'business as usual' lifestyle. In this case the turbine may result in greater energy consumption.

However, human behaviour is difficult to predict and assumptions along these lines may be oversimplifying the situation. There are several issues which should be considered in parallel in order to obtain a more complete view of this situation. For example:

- The 'information age' has brought forth increasing levels of awareness from individuals and perhaps one should not underestimate an occupant's ability to appreciate the essential details. The physical presence of a turbine will in itself generate interest and being a source and stimulus for discussions.
- Many of the users drawn to occupy a building with integrated renewable energy and sustainable design features will be those who will want to work together with these systems in order to lead a sustainable lifestyle. This tendency is likely to increase with time as sustainable living becomes a more relevant issue.
- Historically, individuals who have had energy monitors such as 'smart meters' in their homes (or renewable energy generation) tend to use less energy as their awareness of their energy use grows.

Figure 2.22 The top of the iconic Castle House residential tower, with three integrated 9m diameter HAWTs, in Elephant and Castle. The client/developer is Brookfield Europe and the architects are Hamiltons (Hayes Davidson)

- Occupants in neighbouring areas (or others who have seen these buildings through the media or heard about a project through word of mouth or the internet) may also be encouraged to take some form of positive action (e.g. engage in energy efficiency practices).

Considering these aspects, the overall effect of turbines integrated into the design of buildings could be quite positive, even if the overall percentage of energy supplied is low. Combining this technology with energy-efficient systems, energy monitoring and displaying, sustainable living from occupants, decentralized renewable energy generation nodes, energy recovery and reuse and the complete picture may be inspiring.

Building-integrated turbine design examples

In 2006, planning permission was granted for Castle House – a 43-storey landmark residential building in the centre of Elephant and Castle, London. The design envisages that the top of the building house three 9m diameter turbines in shrouds conceived to accelerate winds and help create more energy.

Usually, when considering the distance between turbines and residential areas, the separation distance is large. However, building-integrated turbines can have a physical barrier between the emission source and the occupant. In the case of Castle House there is the opportunity for a considerable physical barrier between the turbines and the nearest residential unit directly below.

Figure 2.23 The COR tower with integrated HAWTs proposed for Miami's Gold Coast tower (architect Chad Oppenheim)

Several HAWTs have been proposed to be integrated at the top of the 120m COR mixed-use tower, which has been approved by Miami's Design District. This building incorporates photovoltaic panels and solar hot water generation, and is wrapped in a 'hyper-efficient shell'.

Vertical axis wind turbines (VAWTs) can also be integrated into buildings. Project ZED (Towards Zero Emission Urban Development), one of the first concept projects for zero CO_2 emission buildings, was part sponsored by the EU APAS (European Commission DG XII) between 1995 and 1997. It brought together teams from London (Future Systems, BDSP Partnership), Toulouse, France (Ecole D'Architecture), Berlin, Germany (RP+K SOZIETÄT) and Cologne, Germany (TÜV

RHEINLAND). The London team incorporated a large bespoke vertical axis wind turbine in the centre of an aerodynamically shape building designed to maximize wind acceleration while simultaneously demonstrating a high performance building envelope and a visually striking architectural form.

A real-life example of VAWT integration can be found at Technisches Rathaus, Munich, Germany, which has a simple three-bladed H-Darrieus turbine installed on the top of a cylindrical tower. This turbine powers a large slowly rotating external artwork during the day and lighting at night. The rotor was designed by Neuhäuser Windtec GmbH and is rated at 40kW.

Vertical axis wind turbines can be favoured by some purely on the basis of aesthetics. This form of wind energy technology is discussed in depth in Part 4 where comparisons are made between VAWTs and HAWTs. As will be seen, the VAWT in particular can take a variety of forms and can still remain as efficient as HAWTs.

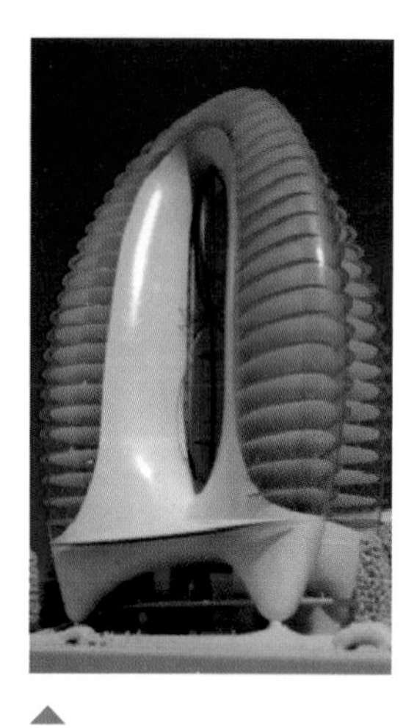

Figure 2.24 Conceptual building-integrated wind turbine from Project ZED (Future Systems & BDSP Partnership)

The Burj al-Taqa (Energy Tower), Dubai, proposed by architectural firm Gerber Architekten international GmbH (and environmental engineers DS-Plan),[12] uses a bespoke VAWT (Patent No. 4-06-05-6331). The 68 storey 322m tower is conceived to be a 100 per cent energy self-sufficient building with 15,000 square metres of building-integrated solar cells. An additional 'island' of solar panels has been proposed to be built adjacent to the building with links to energy storage systems involving hydrogen and hot water. The energy-efficient building envelope has a solar shield and mineral coated 'vacuum glazing'. The fresh air supply is driven by wind forces acting on atria and pre-cooled by sea water.

Figure 2.25 'H-Darrieus type' of VAWT integrated into the design of the Technisches Rathaus, Munich, Germany

The importance of schemes where a developer proactively takes responsibility for more sustainable development practices such as renewable energy integration is also clear in light of specific political agendas, for example, in The Mayor of London's Energy Strategy.[13] The Mayor's targets have called for significant developments to have 10 per cent of the energy needs (power and heat) from on-site renewable energy generation. Measuring the success of the scheme by relating the energy generation capacity to the amount of energy used on-site highlights the need to couple renewable energy generation with energy efficiency.

Figure 2.26 The Burj al-Taqa (The Energy Tower), a 100 per cent energy self-sufficient 68 storey skyscraper with a bespoke VAWT proposed for construction in Dubai. (Gerber Architekten)

4) THE FUTURE OF URBAN WIND ENERGY

The future of wind energy need not necessarily be confined to the types of example already given in this section. The realms of possibilities extend beyond the examples shown in the three main categories presented. Other ideas may be worth considering even if the purpose is only to help remove stereotypical ideas that wind energy should be based on historical and economic factors (rather than being based solely on what is the most appropriate design).

Figure 2.27 The Beacon designed by Mars Barfield Architects which consist of five 'QR5' 5kW VAWT by UK Manufacturers Quiet Revolution

'The Beacon', which was conceived by Marks Barfield Architects and Quietrevolution, incorporates five 'QR5' vertical axis turbines (shown in Figure 2.27) held some 40m above ground level in a sleek silver Y frame. This Y-frame is designed to rotate into the wind and will generate an estimated 50,000kWh/a if sited in areas with annual mean wind speeds of 5.9m/s at hub height. The Beacon, as the name suggests, represent more than a simple means of energy generation. However, an 'energy input – energy return' calculation would be necessary if wide-scale distribution was to be seriously considered. Although the estimated embodied energy payback has been quoted as around 12 months for the turbine itself, when the Y-structure and ancillaries are taken into account this figure will be higher.

Another example of a bold proposal is the 'Aerogenerator' which has been designed by Grimshaw Architects and Windpower Ltd (Figure 2.28). Due to the structural stability of the design of this form of the Darrieus VAWT it could, it has been suggested, be built to a size where it could be rated at 9MW (where the length of one arm would be around 200m) and be located offshore.

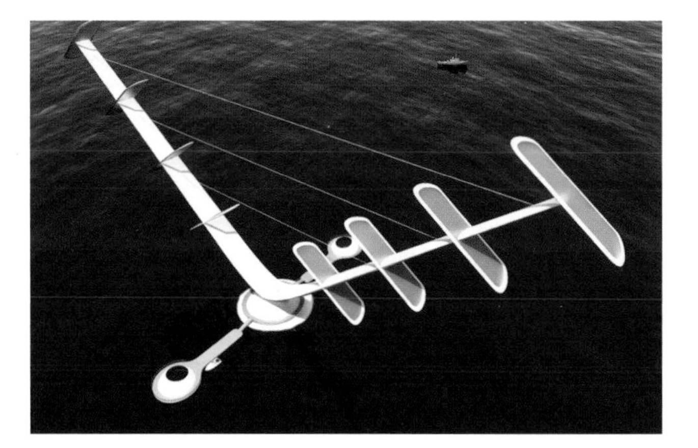

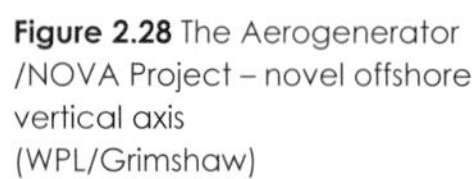

Figure 2.28 The Aerogenerator /NOVA Project – novel offshore vertical axis (WPL/Grimshaw)

Where are we today? A model for the future

Currently, the number of urban wind energy installations is relatively small compared to the potential. However, in some areas, such as the UK, the number are growing rapidly. The BWEA have recorded a growth from 1031 units in 2005 to 13,801 in 2009 for free-standing small wind energy (i.e. less than 50kW). For building-mounted turbines the number of units have risen from 2 to 14,065 over the same period.

The short term future of urban wind energy is uncertain. However, in the long term it could be expected that the increase in momentum of environmental concern, coupled with experience of use and economic factors driven by rising fossil fuel cost, will bring about wide-scale use in areas with suitable wind resources. BWEA predict more than 600,000 small wind energy units will be in use in the UK by 2020 in a market worth over £750 million. By 2040 the number of units is predicted to reach around 4,000,000, generating over 11.1TWh (i.e. meeting 3% of the UK electricty demand).

The development of the large-scale commercial wind farm industry could be viewed as a model for urban wind energy. This development has taken several decades of continuous improvement. Many of these first large-scale turbines suffered from operation and maintenance issues. However, design evolution has not only put an end to the vast majority of these issues but also improved efficiency through improved blade and generator design as well as increasing energy output through increased size. Today both onshore and offshore are well-established and growing industries worth over £35 billion per year.

The economic incentives for wind energy development are in place in many regions to allow this development to occur. However, the urban wind energy market is not as attractive to investors interested in larger projects and higher returns.

In simple terms, if the wind energy resource is identified to be adequate (through on-site monitoring) then an opportunity exists. However, translating wind energy into the urban environment throws up its own issues related to safety, environmental impacts and benefits, cost and payback, reliability, output prediction, energy integration, and performance in complex urban windscapes. Although some of the hard work in turbine design has been done by the development of large-scale turbines, in some respects it is both assisting and limiting the development of urban wind energy. The technology and know-how is in place to get the large-scale turbine up and running next to the end users. However, the bias toward the established design may mean lack of investment in new, perhaps more appropriate design – e.g. VAWT. There is a tendency to consider only what is readily available (and affordable) rather than to design what is most appropriate and develop that technology to the stage where it can be economically viable.

Urban wind energy does have a major advantage, however, in that it can be realized by individuals and communities. It can be carried out in more manageable increments – one energy-generating node at a time. If the economics can be justified, environmental impact mitigated, appropriate equipment selected and good practices adhered to, then the future of urban wind energy can be positive.

Figure 2.29 SkyZED zero energy development proposal for Wandsworth in West London with wind turbines on top and between 35-storey aerodynamic towers of commercial and residential units (Bill Dunster Architects)

As urban wind energy can be intimately linked with building design, its future is tied to the future of architecture. Architecture has always moved through its own evolution process and the modern architectural form has been released to literally new heights by the advances in capability, e.g. via structural design and transportation (lifts). Buildings approaching 1km in height are being built such as 'Burj Dubai' (800m), with a number of developers proposing to exceed 1km (e.g. 'Al Burj', also in Dubai). Complex curved façades, and specialized materials are now practical options – both lending themselves to wind energy integration.

The architectural dictum 'form follows function' is now more relevant than ever with environmental design moving up commercial agendas. The concept of the green tower is emerging, e.g. the 'Tree Tower' conceptualized by William McDonough (author of *Cradle to Cradle* and winner of the Presidential Award for Sustainable Development in 1996). Other offerings for 'tall and green' include Ken Yeang's 'Bioclimatic Skyscraper' and Bill Dunster's SkyZED ('Flower Tower') development (an example proposed for Wandsworth, London, shown in Figure 2.29).

These buildings include a variety of environmental aspects, such as forms that inherently make use of natural ventilation and daylight, and those that enable and enhance local and renewable energy generation.

Emerging technological innovations

While social and political innovation are required to bring forth a new era of renewable energy, the wind industry remains predominantly led by technology. The high level of technological innovation, which has brought about the relatively rapid maturation seen for large-scale turbines over the last decade, is continuing.

Innovation within the manufacturing processes continues to improve reliability while reducing costs. The use of finite-element stress codes are now routinely used to ensure that key components can withstand the huge loads placed on them by weight, motion and 50-year-return wind speeds. The form of these components is analysed to ensure the stresses are evenly distributed. This prevents excessive 'over-designing' and thus reduces the amount of raw materials required in the manufacturing process. It also reduces the weight of the components (which act on other components) and so creates positive knock-on effects. The move to the use of lightweight composites is also contributing to the overall benefits of these advancements.

Generator and tower designs continue to improve (as described in more detail in Part 4). Other potential innovations include 'frictionless bearings' via magnetic levitation, which has the ability to boost wind energy-generating capacities.

The efficiency of blade designs in extracting the kinetic energy from the wind is tending to increase for those companies actively involved in research. Some of the large-scale turbines have active pitch control with a pitch resolution of 0.1 degree and response time of 30ms in order to ensure the blades are at the optimum angle during changing wind speeds. Altering blade design in accordance to new understanding of blade responses to the complex turbulent air structures resulting from different wind conditions is allowing the coefficient of performance to move ever closer towards the theoretical Betz limit of 59 per cent.

Simulating the fundamental physical phenomena associated with wind energy is one of the keys to understanding performance-related issues such as blade interactions with wind. However, huge computational calculations are required to represent reality at the appropriate level of accuracy.

Fortunately, following from Moore's law, the price performance of computational power continues to double every 12 months. If Moore's law continues, and there are indications that this may be possible, in 10 years' time a fixed amount of capital will be able to buy computing power 1000 times greater than today. This is a simple example of the power of exponential developmental growth which many technologies are experiencing.

The developments in computation power have seen computation fluid dynamics (CFD) simulations of aerofoils moving away from 2-dimensional representations to include the third spatial dimension. This is important as air does not merely flow over a blade but also moves in the axial direction. A fourth dimension – time – is now routinely examined to extract data on dynamic flow variations. These hold considerable importance as wind and blade/wind interactions are transient. The sequence of images in Figure 2.30 depicts a '4D moving mesh CFD simulation'. Pressures are plotted on the blade surface and a passive scalar ('smoke') is emitted from one of the blade tips to show the dynamic aspect of the simulation.

Increases in the coefficient of performance C_p from the typical values of around 30 per cent to say 40 per cent are well worth the effort as this corresponds to a 33 per cent increase of energy output 'for free'. Part 4 expands on C_p and shows some of the highest performing turbines.

Airflow optimization not only applies to blade design but also building and shroud forms. The technology is in place to perform this optimization automatically by coupling CFD codes with optimization codes which use genetic algorithms (GA) to optimize a number of parameters simultaneously. This generally involves simulating several hundred cases; however, the continuing advances in computational hardware bode well for the future of blade and building optimization.

Figure 2.30 Transient 'moving mesh' CFD analysis of a small turbine, revealing pressures and lift forces on blade surfaces and modelling smoke emission from a blade tip in order to show the time dependency of these models (BDSP Partnership)

SUMMARY

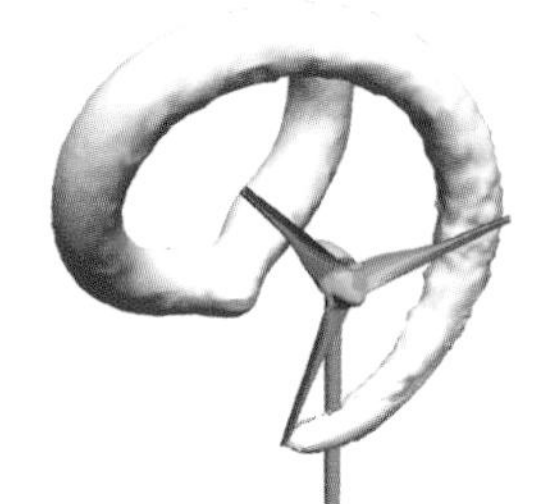

The potential for urban wind energy in windy urban areas is large and there are many forms that can be exploited. The various examples presented point to this significant potential of urban wind turbines to meet energy needs as part of a comprehensive diverse energy portfolio (based on the local environmental design opportunities).

Depending on the situation in hand, each of the three main wind energy categories will vary in appropriateness. For individual homeowners and enthusiasts, there is potential for home-mounted turbines. A familiarity with the basic concepts of wind energy is required. In addition, wind energy resources should be assessed/monitored and the turbines should be sufficiently elevated.

For those wishing to come together in groups, large-scale stand-alone turbines may be an attractive option worth considering in terms of community ownership and investment. Similarly, developers and investors wishing to provide considerable amounts of energy may wish to consider a large stand-alone turbine – e.g. to meet the electricity demand of 500 homes.

For those involved with urban redevelopment and concerned in moving our culture forward in a sustainable direction, building-integrated turbines, where the buildings are developed with wind energy in mind, may be worthy of investigation.

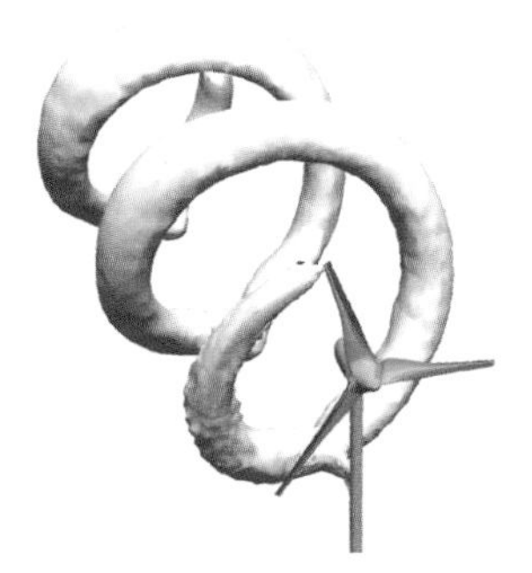

Architecture is one of the higher forms of human expressionism and in certain circumstances is able to reflect aspirations of a generation and become a source of inspiration. Urban wind energy is part of an environmental led design approach where form follows function. Environmental design is also concerned with human comfort therefore wind energy should be designed so as not be to the detriment of wellbeing (which includes aesthetics).

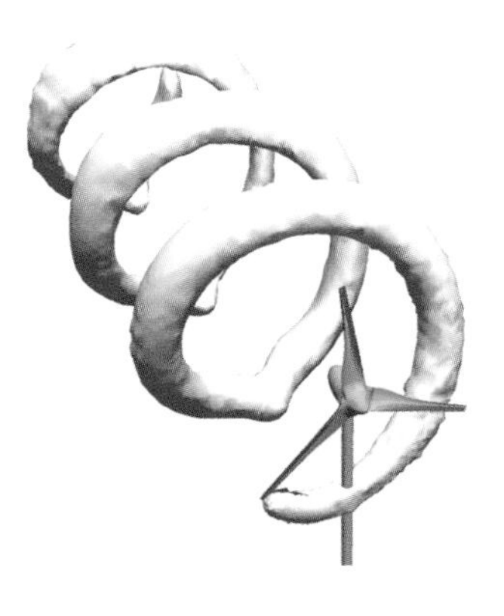

Creativity is required when engaging in these types of projects and there may be a tendency towards more organic, curved shapes. As wind energy, by virtue of it highly visible nature, draws attention it can be used to create iconic forms (and can be thought of in terms of an 'active logo' or 'dynamic art').

Wind energy is compatible with other electricity-generating technologies (e.g. PV systems and CHP) and should be part of a sustainable design philosophy which also promotes sustainable living habits.

The future of urban wind energy and the extent to which it will be changing urban landscapes is unclear. Urban wind energy is in its infancy; however, the number of projects is growing and the drivers are continuing to gain weight. The instigators and supporters of these projects have helped to establish precedents which, after each project, make the pathway a little easier for others to follow. Further details of some of these projects are included in the following parts. As these projects are only the beginnings of ventures into this arena, they largely err on the side of caution and so there may be scope for new developments to push boundaries forward.

The potential of the visible aspect of wind energy to influence individuals should not be undervalued. Whether via an iconic stand-alone or fully integrated wind energy installation, if the motives are genuine, a clear message is being sent out. In simple terms these messages are:

- We think there is a serious problem (relating to energy/ sustainability) that needs addressing.
- We care enough to act / We are doing our part.
- We want set examples and help others do the same.

The presence of one positive social input, for example, renewable energy generation, can help bring forth others and may encourage:

- energy monitoring;
- sustainable living (not only for energy use, but for transport, local foods etc);
- installation or use of energy efficient systems such as A-rated white goods, or additional insulation;
- linking to other building and external entities – e.g. linking to local energy generation nodes and waste heat/district heating schemes.

The next section moves from the initial 'cause for action' and 'conceptual' phases to 'planting feet on solid ground' where practical aspects such as initial actions, decision-making and project feasibility are considered.

REFERENCES

1 www.bwea.com/small/ (September 2008)

2 www.urban-wind.org/index.php (September 2008)

3 P. Gipe, *Wind Power*, James & James (2004)

4 Beama Energy DTI sponsored project; Reference: K/EL/00312/00/00, http://83.217.99.100/cfide/beamaprj/downloads.cfm (July 2008)

5 www.warwickwindtrials.org.uk (July 2004)

6 Hockerton Housing Project Trading Ltd., The Watershed, Gables Drive, Hockerton, Southwell, UK. www.hockertonhousingproject.org.uk (September 2008)

7 Steve Rose, *The Guardian* (18 July 2005)

8 www.london.gov.uk/mayor/planning_decisions/strategic_dev/2003/aug1303/dagenham_ford_wind_turbines_report_barking_and_dagenham.pdf (September 2008)

9 S. Merton, *Wind Energy in the Built Environment*, Multi-Science (2006)

10 N.S. Campbell and S. Stankovic, *Wind Energy for the Built Environment - Project WEB*, published by BDSP Partnership Ltd on behalf of the Project WEB Partners (BDSP Partnership Ltd.; Department of Aeronautics, Imperial College, London; Mecal Applied Mechanics BV.; Institut fur Baukonstuktion und Entwerfen L2, Universitat Stuttgart; CRCL Rutherford Appleton Laborotory; Xkwadraat) (2001)

11 http://atkins-me.com/Project.aspx (June 2008)

12 www.gerberarchitekten.de/englisch/projekte/745/745.htm (September 2008)

13 Mayor of London, *Green Light to Clean Power: The Major's Energy Strategy*, www.london.gov.uk/mayor/strategies/energy/docs/energy_strategy04.pdf (February 2004)

Urban Wind Energy Feasibility Study

'Tulipo' turbine at Knivsta High School in Sweden, which has a 5m blade diameter and is rated at 2.5kW)
(WES – Wind Energy Solutions)

INTRODUCTION

INITIAL INVESTIGATIONS, DECISION-MAKING AND THE FEASIBILITY STUDY

Once ideas and concepts have been proposed and shaped into potentially attractive and workable forms preliminary investigations into viability can begin. These are usually brought together in a concise yet comprehensive feasibility study. This text will inform the initial decision-making process from which the go-ahead may be given to proceed to the next level of design. This stage will require a higher level of commitment. It may involve on-site wind resource or acoustic monitoring and establishing contact with planners and key stakeholders. Manufacturers will also be approached for their inputs and information such as budget quotes, performance data, noise emission data and lead-in (delivery) times. A full 'feasibility study' will be necessary for all large-scale projects and is often carried out for projects as small as 5kW when in an urban setting.

It is important that those involved from the outset have a good understanding of wind energy fundamentals in order to avoid basic pitfalls before going to the first level of commitment, i.e. commissioning a full feasibility study.

The feasibility study should address, as a minimum:

1. Project aims
2. Initial wind resource estimation and site study
3. Environmental impacts and suitable/available technologies
4. Economic aspects (which include energy yield estimations, and operation and maintenance issues)

It may also contain: overall recommendations; advice on the planning process and available grants; examples of similar existing projects and community involvement strategies; etc. For large-scale urban wind energy projects, addressing genuine community concerns is vital and can help attain planning permission. This may involve promoting several aspects beneficial to local communities such as education, community identity and funding community projects with generated income. In general, communities are more receptive (and tolerant to any impacts) if they can understand or perceive or, better still, share in the benefits.

When progressing an urban wind energy project through the planning permission application stage (local, regional or national), a formal feasibility study (or elements of the study) can be submitted directly as part of the application. Larger projects may even require a formal environmental impact assessment (EIA), although this is not generally required unless a wind energy project is a commercial development of five or more turbines or over 5MW in capacity.[1]

This section expands on each of the four main areas listed above in terms of general urban wind energy.

1) PROJECT AIMS

A formal statement (or brief) setting out the aims and the specifics regarding the context or spirit in which a particular project is being approached can be very useful with respect to the overall decisions that will be made from the results of a feasibility study.

If the aims of a given project are clear and well conceived, it will be easier to keep in mind the most important aspects of the project which can get lost when generating and processing the finer details.

The ideals set out in this section of the feasibility study may also be useful for others, such as the planners or stakeholders, who will decide how their values align with the proposal and to what degree to offer their support . It can also help add momentum to a project by reminding the various actors why renewable energy may be an integral element of a larger project vision.

Project aims, for example, may emphasize the importance of capital returns for reinvestment into a community and the sustainable running of the scheme (i.e. to pay for operation and maintenance). For certain projects, the educational and demonstration properties may be given elevated status (e.g. for turbines located within the grounds of educational facilities). Other projects may wish to emphasize the goal of adding value to a development in terms of increasing the iconic status and visibility of sustainable design aspects which may be important to the client or potential property leaseholders.

Developers wishing to strongly address the issue of energy consumption of a development may want to highlight the aim of linking on-site energy generation with, for example, smart metering in order to bring energy awareness to the end users of associated buildings. Some teams may want to stress the importance of 'future proofing', i.e. integrating a project with emerging future developments such as local energy networks. In short, the creation and effective communication of a clear and manifestable vision is encouraged.

2) INITIAL WIND RESOURCE ESTIMATION AND SITE STUDY

Assessing on-site wind resources in the appropriate manner is vital for any turbine feasibility assessment and its importance cannot be emphasized too strongly. A considerable over-estimation of wind resources can be disastrous for an installation as the available energy for harvesting is very sensitive to the available wind speeds, as discussed.

In the case of turbines to be sited in the built environment the likelihood of this undesirable state occurring is increased due to the complex nature of both urban geometries and the associated wind flow patterns which tend to reduce available energy resources, as well as presenting additional problems related to increased turbulence, as discussed later. Conversely, under-predicting resources may mean that opportunities for successful and beneficial wind energy installations may be missed.

It should also be noted that wind turbines, at least larger models, are optimized for different IEC wind classes (see Table 4.2, page 136), so inadequate wind data can also lead to inappropriate equipment selection and lower energy capture.

The available wind resources will depend primarily on large macro-scale (continental) and meso-scale (regional) conditions. However, in the urban environment, micro-scale (local) circumstances can be critical.

Although on-site monitoring is the best way to determine available wind resources, macroscale wind resources can be initially assessed by recourse to data from local weather stations, and micro-scale effects can be initially appreciated through qualitative inspection.

Before examining these two aspects of the wind energy resource in detail, a brief summary of wind speeds in relation to wind turbine use is given in Box 3.1.

BOX 3.1

SPECIFIC WIND SPEEDS IN RELATION TO WIND TURBINE OPERATION

Cut-in wind speed: At this speed the wind provides enough force to begin to turn the blades. The value depends on the blade design and the friction-generating elements of the drive train.

Start up wind speed: At this speed the blades are moving fast enough, and are capable of transferring enough torque to the drive shaft, to enable useful electricity to be generated. At this speed the generator will start to operate and produce useful electricity. In the case of a Darrieus type VAWT, which does not have the capacity to self-start, this is the wind speed at which the starter motor sets the blades in motion. Although the start-up wind speed can be very close to the cut-in speed they are not the same. For example, the Bergey XL1 turbine has a cut-in speed of 2.5m/s; however, the start-up wind speed is just over 3m/s. Although the turbine may be able to generate some electricity at 2.5m/s it may not be compatible with the electrical grid.

Minimum annual mean wind speed: The annual mean wind speed is simply the wind speed at a certain location averaged over a year. If the annual mean wind speed at a site is equal or greater than the designated 'minimum annual mean wind speed', there will be enough energy in the wind on an average basis to begin to consider the idea of installing a wind turbine on the corresponding site. Although, there is no definite point where the technology will move from unfeasible to feasible, a useful value to keep in mind is a minimum annual mean wind speed of 5.5m/s. It should be noted that this refers to the average speed of the wind at the turbine hub height which could be, for example, 70m or more and not the general site speed, which may have been taken from a standard weather station with an anemometer 10m above ground level.

Rated wind speed: This corresponds to the maximum energy the turbine can extract from the wind. The rated wind speed is sometimes referred to as the 'name plate value' as this is the peak value quoted when referring to a particular turbine. Beyond this wind speed the turbine will either passively or actively reduce the percentage of energy it will extract from the wind in order to prevent damage to the device.

Cut-out wind speed: At this speed the wind turbine will stop turning completely in order to prevent damage to the turbine. This cut-out speed is usually quite high, such as 25m/s, and will rarely occur on most sites.

One other wind speed term to consider is the storm-rated wind speed (or survival wind speed). This is the wind speed that a given wind turbine is known to be able to withstand without damage (e.g. 60m/s) and can vary significantly depending on the turbine in question. This can be critical for an urban wind turbine if winds are being deliberately accelerated.

Table 2.1 Typical values for key wind speeds

	Wind speed					
	m/s	km/h	mph	ft/min	ft/s	knots
	1	3.60	2.24	196.85	3.28	1.94
	2	7.2	4.5	394	6.6	3.9
Possible cut-in speed	3	10.8	6.7	591	9.8	5.8
Typical cut-in speed	4	14.4	8.9	787	13.1	7.8
	5	18.0	11.2	984	16.4	9.7
Minimum annual mean speed	6	21.6	13.4	1181	19.7	11.7
	7	25.2	15.7	1378	23.0	13.6
	8	28.8	17.9	1575	26.2	15.5
	9	32.4	20.1	1772	29.5	17.5
	10	36.0	22.4	1969	32.8	19.4
	11	39.6	24.6	2165	36.1	21.4
Typical rated speed	12	43.2	26.8	2362	39.4	23.3
	13	46.8	29.1	2559	42.7	25.3
	14	50.4	31.3	2756	45.9	27.2
	15	54.0	33.6	2953	49.2	29.1
	16	57.6	35.8	3150	52.5	31.1
	17	61.2	38.0	3346	55.8	33.0
	18	64.8	40.3	3543	59.1	35.0
	19	68.4	42.5	3740	62.3	36.9
	20	72.0	44.7	3937	65.6	38.9
	21	75.6	47.0	4134	68.9	40.8
	22	79.2	49.2	4331	72.2	42.7
	23	82.8	51.4	4528	75.5	44.7
	24	86.4	53.7	4724	78.7	46.6
Typical cut-out speed	25	90.0	55.9	4921	82.0	48.6

Macro-scale wind speeds

The general wind character of a region of a particular country has three components that are key to wind energy resource assessment. The main factor, the 'annual mean wind speed', has been mentioned already. This is available from local weather stations and is equal to the sum of the hourly average values for the whole year divided by 8760 (the number of hours in a year). The mean wind speed will depend on many factors such as the location in relation to dominating global wind currents, the distance from the coast (i.e. the amount of upstream 'roughness' that the winds have to travel over to reach the site as well as other conditions such as the altitude of the site.

The second factor is the 'wind distribution profile'. Information on the distribution of the wind speeds (i.e. how many hours a year the wind will be calm, 0.5m/s, 1m/s, 1.5m/s, 2m/s etc) allows the energy content of the wind resource of a specific site to be more accurately calculated. This is important as the energy available in the wind is a cubic function of the velocity and therefore two regions with the same annual mean wind speed could have very different total annual energy contents.

Wind frequency data are often presented using standard statistical functions. The 'Weibull distribution' is the most common and can be presented in two forms: the 'probability density function' and the 'cumulative distribution function'. The 'probability density function' produces the most intuitive results and so is most commonly used. In this case, the velocity distribution is a function of k, the Weibull shape factor, and c, the Weibull scale factor. The full form of the Weibull velocity distribution equation is given below:

$$f(V) = \frac{k}{c}\left(\frac{V}{c}\right)^{k-1} e^{-\left(V/c^{\,k}\right)}$$

Two idealized Weibull profiles are presented in Figure 3.1. The most common sub-set of the Weibull profiles is the 'Rayleigh distribution' where k = 2m/s. Although the mean wind speeds are the same for these two idealized wind distribution profiles the total available energy is greater in the case of the Rayleigh distribution (25 per cent more energy).

If energy yields are estimated purely on annual mean wind speeds alone (i.e. assuming the wind is constantly at the mean wind speed) the energy yield would be underpredicted by a factor of 1.91 (if the data for the wind distribution profile fitted the Raleigh distribution). Therefore, as a simple approximation, the energy yield can be calculated on the basis of the mean wind speed and multiplied by 1.91 assuming the Rayleigh distribution is valid for the site in question (which may be an acceptable preliminary guess in the temporary absence of actual wind data).

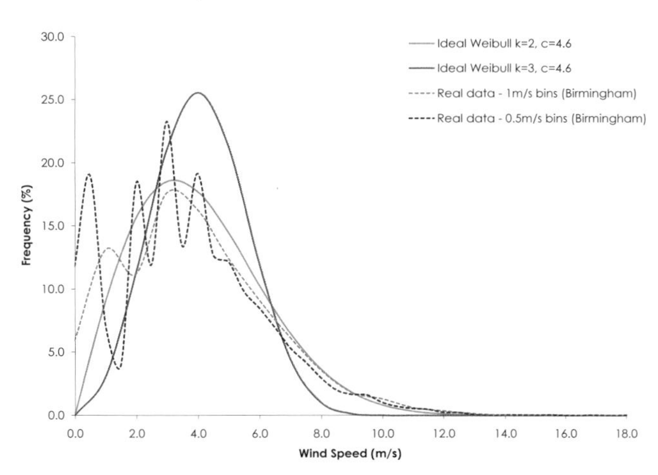

Figure 3.1 Wind speed distribution from ideal statistical representations and real data

In addition to the two idealized wind distribution profiles, Figure 3.1 presents some real wind data (which originate from a standard weather station in Birmingham, UK). As can be seen, the real measured data usually differ from the idealized case (from which manufacturers produce their energy yield estimates). When the real data are grouped into 1m/s 'bins' the ideal Weibull (k = 2 and c = 4.6) fits reasonably well. However, when the more accurate 0.5m/s 'bins' are used the extent of the discrepancy between real and ideal can be seen.

The 'cumulative distribution function' can be used for estimating the fraction of time T the turbine will be turning. This may be a very important consideration if a turbine is to feature in a prominent position in the urban landscape or be considered iconic. In the case of a cut-in speed of 4m/s and cut-out speed of 25m/s we can use the following formula:

$$\mathbf{T} = e^{-\left(4/c\right)^{k}} - e^{-\left(25/c\right)^{k}}$$

Using the above formula for a Rayleigh distribution where k = 2 and the shape factor c = 4.6 we can see that the turbine will only be turning and generating some electricity for about half of the time:

$$\mathbf{T} = e^{-\left(4/4.6\right)^2} - e^{-\left(25/4.6\right)^2} = \mathbf{0.47}$$

The third component is the wind direction. Although this may not be a major concern for wind farms developed in open-field areas, it can prove very important when considering urban wind energy and assessing the suitability of a particular location.

Analysing the average energy content of the wind from the various wind directions gives an appreciation of where the dominant high energy flows originate. This information can be related to the proposed location to determine an optimal turbine position or, at least, to avoid obstacles such as tall buildings and trees upstream of the turbine.

In the case of building-integrated turbines, knowledge of directionality can be even more important as the buildings will often be designed to accelerate winds from certain key directions. The subject of wind directionality and classification will be discussed in Part 5.

Micro-scale wind speeds

Figure 3.2 illustrates how the differing arrangement of simple buildings can disturb the wind flow, generate varying wake patterns and induce swirling turbulent flow. When siting turbines in an urban environment these disturbed flow zones should be identified and avoided. This is commonly achieved by ensuring the blades of the turbine are sufficiently elevated above roof level. Generally, the disturbed flow region in the 'isolated roughness flow' case is considered to be twice the height of the obstacles. Therefore turbine blades, generally, should be located twice the height of the tallest local obstacle to avoid a significant drop in potential performance.

There are cases where the acceleration near buildings can be used to gain an advantage but usually if the building has not been carefully designed with wind energy in mind this should be avoided. The wake region in the isolated roughness case is considered to extend to between 10 and 20 times the obstacle height.

Figure 3.2 Urban wind regimes showing 'disturbed' regions around simple buildings characterized by lower velocities and high levels of turbulence (adapted from Oke[2])

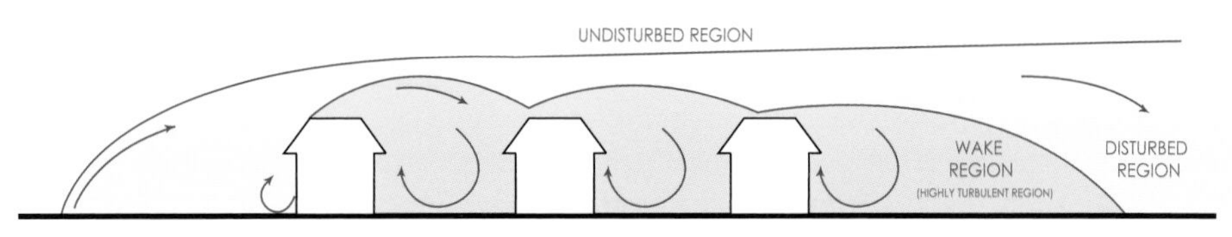

SKIMMING FLOW

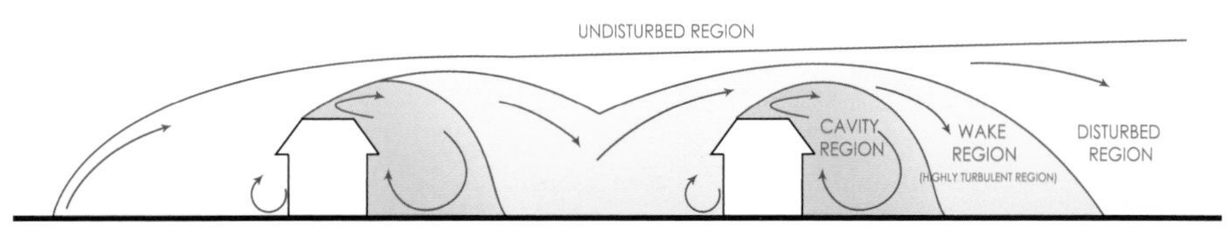

WAKE INTERFERENCE

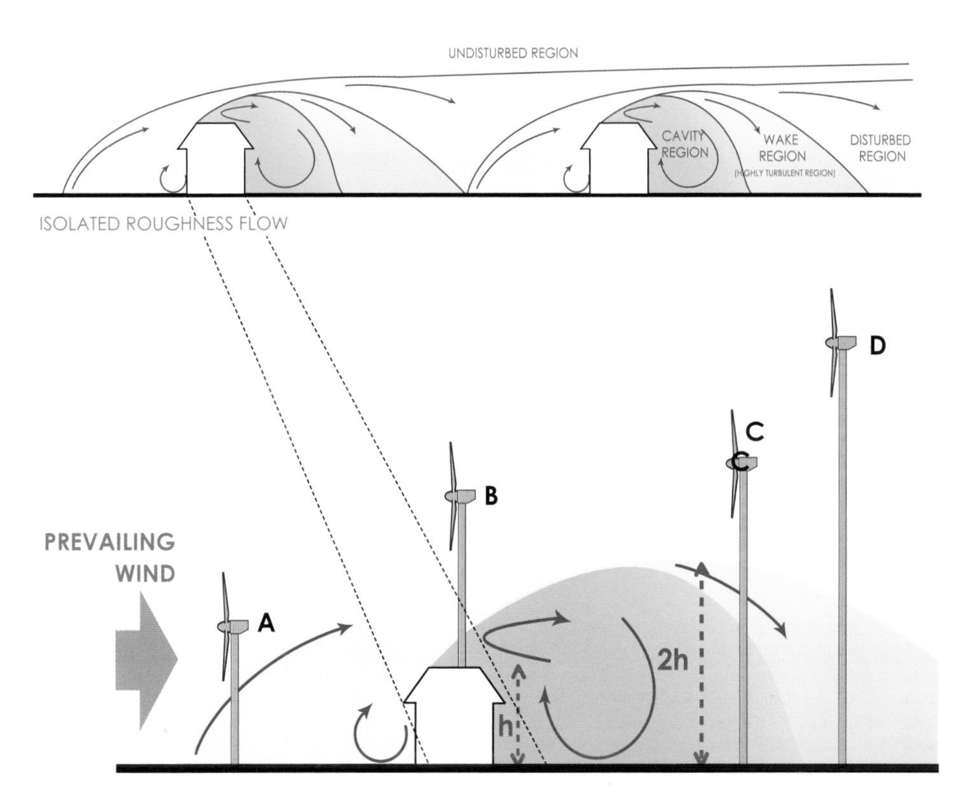

Turbine **A** is at a low height and still avoids the highly turbulent region (provided the wind blows in the direction of the prevailing wind)

Turbine **B**, mounted on the top of the building, is elevated into the undisturbed region

Turbine **C**, located above the wake region, requires the blades to remain above a minimum height of 2h

Turbine **D** is positioned well above the disturbed region and exposed to the higher wind speeds generally available at higher elevations

When dealing with urban winds, areas separated by very small distances, as little as a few metres in some cases, can experience very different annual wind speeds. Again, the way to avoid these complex fluctuating winds is to elevate a particular turbine into the 'undisturbed region'. When evaluating a location for a turbine, the character of the immediate area and neighbouring regions related to the size and density of the obstacles should be assessed. These are evaluated via 'terrain roughness' categories.

As there is friction between the wind and the ground, velocity gradients (wind shear) develop, with the best wind resources (higher wind speeds) occurring at greater heights. The extent of these variations depends on the local obstacles, or terrain roughness of the immediate surrounding area and also the roughness upstream of the site.

In the example below, three separate terrains are shown with their corresponding 'aerodynamic roughness': city centre terrain (Z_o>0.7), a suburban terrain (Z_o = 0.25–0.3) and an open-field terrain (Z_o = 0.01–0.03). For very aerodynamically rough areas, the profiles require the use of a displacement length, d, to show how the profile is displaced relating to the height and density of the buildings. At 30m above ground level the wind speeds in a city are much lower than in an open field. Furthermore, in this case, a turbine would have to be placed at 70m above ground level to experience the same wind speeds as an open field would have at 60m above ground level. It should be noted that, depending on the degree and area of roughness, accessible winds in urban areas may never reach the same speeds accessible at much lower heights in a nearby open-field.

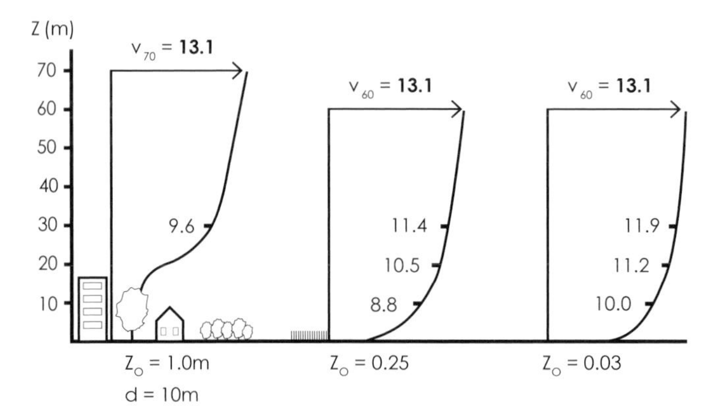

Figure 3.3 Indicative wind speed profiles associated with different terrains (and their corresponding aerodynamic roughness, Z_o, and displacement height, d)

What, where and when to assess?

Wind resource assessment can be carried out in several stages. The first stage takes place during the initial feasibility study and involves estimating local wind speeds using third party sources. These not only include local weather stations but also macro-scale mathematical models, for example, the numerical objective analysis of boundary layer (NOABL) wind speed database for the UK.

The NOABL database is made freely available by the department of Business Enterprise and Regulatory Reform (BERR)[3] (since 2000) and provides an estimation of average annual wind speeds across the UK resolved to one square kilometre. The data are the result of an airflow model that estimates the effect of topography on wind speed and is specifically designed for wind turbine project feasibility studies. The average annual wind speed data are given at three heights above ground level (10m, 25m and 45m). It should be noted that the NOABL database is designed for preliminary assessments only and is known to contain errors. For example, it does not take into account terrain roughness changes. The terrain roughness of the site and the upstream terrain changes are important as the rougher the terrain (i.e. the higher/denser the obstacles such as trees and buildings) the steeper the velocity gradients and the lower the wind speed at a given height. Therefore, one may expect to encounter lower wind speeds in reality than indicated by NOABL database and one group reporting on wind data taken from building-mounted wind turbine sites in urban areas indicates that wind speeds may be around 20 per cent lower than those predicted.[4]

Steeper velocity gradients are also likely to exist in urban areas which will mean that tower heights (i.e. the height of the hub above the ground level) will often be a very important factor. This is evidenced from NOABL scaling factors proposed by Encraft from the results of the Warwick Wind Trials, which range from 1 (for turbines mounted 5m above the ridgeline on top of high-rise buildings that are over 10m higher than surrounding buildings within a radius of 10m) to 0.4 (turbines mounted where there are buildings and trees within a 10m radius that are higher than the turbine hub height).

As a basic guideline, annual mean wind speeds of 5.5m/s and above can be considered for a wind energy installation. The National Renewable Energy Laboratory (NREL), which provides

wind data for many countries,[5] classifies mean wind speeds (at 50m above ground level) below this value to be 'poor'.

A weather station will be able to provide data at various levels of detail. However, this will usually be made available at increasing levels of cost. The annual mean wind speed for a 10-year history can be requested or, for example, the data presented in the form of a 'wind rose'. It should be noted that this information is usually taken at 10m above ground level and will usually be for open-field terrain (although there may be local obstacles impairing the wind from certain directions).

Knowledge of prevailing wind directions combined with an inspection of the local and surrounding terrains may give a good qualitative feel for the viability of a certain project, i.e. if there is considerable roughness upstream of a site for the dominant wind directions then the site should be avoided. Wind roses can be superimposed onto a potential site map to ensure, wherever possible there are 'clear fetches' upstream of the turbine. Wind roses can come in several forms and can provide information on, for example:

- frequency of occurrence of wind from each direction (e.g. the total hours the wind blows from each direction);
- proportion of normalized 'mean wind speed' or more usefully the 'available energy' in the wind for each direction;
- distribution of wind speed magnitude (displayed in discrete bands) for each wind direction.

Should initial resource indications prove favourable (and other factors such as economic viability), the next level of wind data analysis can be approached: on-site data measurement.

Before proceeding, it can be worthwhile to check for availability of wind speed information derived from any other neighbouring wind turbine assessments or installations. Local planning authorities will be a source of potential wind monitoring and turbine locations.

Wind monitoring equipment is usually inexpensive. For wind speed measurement a simple cup anemometer can be used. However, many anemometers come with wind vanes to correlate wind speed with wind direction (Figure 3.4).

Figure 3.4 Combined cup anemometer and weather vane (Ultimeter Pro)

The majority of the costs come from the man-hours involved to carry out the monitoring and from erecting a mast (and if planning permission is required for the monitoring mast).

Monitoring should ideally be carried out for over a year in the exact location of the turbine. The time averaging period should be lower than that found in meteorological stations (1 hour) in order to avoid underestimating the energy content in the air. Ten minute intervals should suffice (time intervals much lower than this may produce too many data sets if data loggers have small memories).

Note: large turbines will pitch their blades into the wind in order to start turning when the wind speeds are above a certain threshold for a specified length of time (e.g. 4m/s for 10 minutes). This can be taken into account when analysing the data to predict annual energy yields. Also, the response time in terms of wind direction can limit the harvesting of certain winds. This only applies to HAWTs and not VAWTs which are always facing the right direction.

'Wind turbulence' is the third parameter that requires consideration when monitoring the wind in urban areas. Although high levels of turbulence relate to increased energy in the wind, the randomness in the direction of the swirling turbulent eddies and gusts works against the action of a turbine.

Typically, high quality winds (with high wind speeds and a low turbulence intensity of say 10 per cent) will, after passing over obstacles like buildings become more turbulent. This not only reduces the amount of energy that can be extracted from the wind but also increases stress on the turbine and associated equipment wear.

Figure 3.5 The CSAT3 sonic anemometer which can also measure turbulence (Campbell Scientific)

Turbulence is not usually assessed for projects outside of the urban environment or for smaller projects such as home-mounted wind turbines due to the additional cost. Sonic anemometers, which can be used to assess turbulence (as well as wind speed and direction), are much more expensive and require a greater data logging capacity.

The variability of the wind on a yearly basis can be quite significant. Therefore, data produced from on-site monitoring should ideally be compared to long-term weather data from

local weather stations. From this comparison the relative intensity of the wind from the data collected can be discerned. This can improve the confidence in the quality of the available local wind energy resources.

For homeowners wishing to investigate their local wind resources, data from inexpensive weather stations can be sent directly to a computer. This bypasses the need for a secure data logger. However, there will be limitations in the allowable cable length between weather station and computer (due to the low voltage).

The main drawback to wind monitoring is the time it takes to collect a large enough data sample. Short cuts can be taken if hourly data if available from a nearby weather station; however, it is not recommended. Desktop wind studies (CFD simulation) provide the quick and relatively cheap means of assessing local (micro) wind environments (this technology is demonstrated more in Part 5). Wind monitoring activities can run in parallel with the planning application. These applications have taken several years in some of the pioneering cases shown in Part 2 although the processing times are much less today and are tending to decrease.

The final site variables that should be estimated are the mean annual temperature and the site altitude. These variables can be used to determine the air density which has an impact on the energy yield (see Box 3.2).

The concluding activity of a wind resource investigation will be to predict total annual energy generation for a particular turbine at a given location. Other aspects can then be evaluated such as CO_2, NO_x and SO_x emissions saved and the percentage of the year the turbines will be turning.

In the spirit in which renewable energy should be approached, the results from any wind monitoring should ideally be made freely available to others (this also applies if the results show a particular site to have poor resources).

BOX 3.2

AIR TEMPERATURE AND DENSITY IN RELATION TO THE AVAILABLE WIND ENERGY RESOURCE

As seen from Box 2.1, the air density ρ is a variable taken into account when calculating energy yields when using the well-known wind turbine power equation (where C_p is the turbine performance coefficient, A is the swept area of the blades and v is the free wind speed):

$$P_{turb} = C_p \tfrac{1}{2} \rho A v^3$$

The air density is usually set as 1.2kg/m³ for initial energy production estimates. However, air density can be an important variable if the annual mean air temperatures are particularly cold or warm, as air density is a function of air temperature (as can be seen from the table below).

Air density as as a function of annual mean air temperature													
Air temperature (°C)	-8	-4	0	4	8	12	16	20	24	28	32	36	40
Air density (kg/m3)	1.33	1.31	1.29	1.27	1.26	1.24	1.22	1.20	1.19	1.17	1.16	1.14	1.13
Energy yield % change*	10.56	8.92	7.32	5.77	4.27	2.81	1.38	0.00	-1.35	-2.66	-3.93	-5.18	-6.39

(*with respect to 1.2kg/m³ i.e. 20°C)

It should be noted that the majority of energy extracted from a turbine may occur during a particular time of year, e.g. winter when the temperatures are cooler. This can also be taken into account with more detailed calculations but will usually not give more than an extra 5 per cent on predicted energy yields. The density of the air also depends on the air pressure. Therefore, the height of the turbine above sea level can also be considered (as the air becomes thinner at higher altitudes). The table below can be used in addition to the table above to take into account the height above sea level of a certain wind energy proposal.

Air density as a function of height above sea level										
Height (metres above sea level)	0	100	200	300	400	600	800	1000	1500	2000
Air pressure (kPa)	101.3	100.1	98.9	97.7	96.6	94.3	92.1	89.9	84.8	79.9
Air temp (°K)	288.1	287.5	286.8	286.2	285.5	284.2	282.9	281.6	278.4	275.1
Air density (kg/m³)	1.22	1.21	1.20	1.19	1.17	1.15	1.13	1.11	1.06	1.01
Energy yield % change*	0.00	-0.96	-1.90	-2.84	-3.76	-5.59	-7.38	-9.13	-13.36	-17.39

(*with respect to sea level)

This table takes into account the decreasing pressure of the air with increased elevation and the temperature lapse rate which is -0.65°C every 100m. It should be noted that air density decreases with increased humidity. This counter-intuitive relationship is due to the fact that the molar mass of water vapour (gaseous H_2O) is lighter than air (e.g. gaseous N_2 and O_2). Hence it may be worth considering the humidity if the region is particularly dry or moist.

3) ENVIRONMENTAL IMPACTS AND SUITABLE/AVAILABLE TECHNOLOGIES

Environmental impacts are of considerable importance when dealing with urban wind energy. A given development should not benefit the global environment to the detriment of the local environment.

The various environmental impacts of any scheme must therefore be properly assessed and measures taken if and where appropriate to avoid unfavourable effects either to the equipment, the immediate surroundings or to the various stakeholders. If the impacts are not assessed in the appropriate manner and mitigating remedies cannot be found for any negative effects that do arise, the local authorities could call for the removal of the device.

This section examines issues relating to:

- public safety;
- visual effects;
- noise;
- shadow flicker and blade-reflected light;
- electromagnetic interference (EMI);
- biodiversity and birds;
- property values/house prices.

Figure 3.6 Large turbines located within fallover distance of buildings in Hoethe, Westerwald, Germany (Paul Gipe)

Public safety

The public safety implications of wind turbine implementation are the first issue to be considered. These are a particularly important aspect for a planning submission. As with all developments, the risks will have to be limited to a quantifiable, generally accepted risk level. For a scheme involving the integration of wind turbines, safety during both construction (i.e. construction workers) and operation (including maintenance) must be addressed.

General public safety risks could include the following:

- major failure of turbine tower and subsequent collapse of the nacelle and blades;
- shedding of (parts of) a blade during operation;
- ice forming and being thrown off the blades during winter.

Examples of turbine failure and even tower collapse can be found in the history of turbine development. However, it may be fair to say that these have generally been as a result of extremely windy conditions or poorly designed installations. Generally urban areas are not associated with extremes in wind conditions. Also, 20 years of design experience (for example with tower designs and foundations) mean many modern turbines present very low risks.

This is reflected in the fact that there are now many examples of even very large turbines integrated very close to buildings and main roads (closer than the fall-over distance) such as those in Green Park (Figure 2.9), Leonardo Da Vinci School (Figure 2.10) and the example in Westerwald, Germany, given in Figure 3.6.

However, although risks can be mitigated, given the current level of inspection and increasing number of installations, technical failures are an inevitable part of wind energy, as can be seen in Figure 3.7.

Figure 3.7 Turbine malfunction in Herenveen in Holland

Over the last few decades, worldwide, there have been several deaths related to wind turbines. The majority are involved with accidents during construction and maintenance. The two public deaths cited by Gipe involve a crop-duster pilot in Texas who struck a guy wire on a meteorological mast and a female parachutist who drifted into a large turbine in Denmark on her first solo jump.[6]

Although the likelihood of a major tower failure over the course of the lifetime of a well-designed turbine is extremely small, it may be the 'perception' of safety that plays an important role in the minds of the public and indeed planners. For example, small or medium turbine towers are commonly supported by the more cost-effective 'guyed' tower types (secured using cables or guys) or 'lattice' type as shown in Figure 3.8. Although these types have proved to be effective in the past, a sturdier turbine tower will give rise to greater 'perception' of safety for people and property in the immediate area.

From a safety point of view, the minimum separation distance between a turbine and the nearest building (or potentially 'sensitive' location) can be thought of as the turbine height plus 10 per cent (i.e. the fall-over distance). This guideline is for large turbines. There are of course many examples of turbines installed on roofs that do not adhere to these recommendations. For turbines located in areas prone to severe storms and hurricanes the distance to the nearest property becomes a more pertinent issue. However, in these areas general damage to property in extremely windy conditions could be greater than that to wind turbines.

Ice build-up will not be an issue in many areas. For example, in England the weather conditions for ice to build up on wind turbine blades occur less than one day per year.[7] As urban turbines are subjected to the 'heat island' effect, the likelihood of ice build-up occurring on turbine blades is diminished

Figure 3.8 Lattice towers from Gorgonio Wind Farm, near Palm Springs, CA (Jim Kolmus)

further. Should ice build-up be viewed as a potential problem in relatively cold climates the maintenance schedule should include monitoring the blades in peak winter to record any occurrence of ice build-up and discharge. In some turbines, ice build-up can cause an imbalance in the blades and trigger an automatic shut-down. Ice build-up will usually collect and fall in wafer thin slices and break up in descent (usually landing near the base of the tower.[8] However, in extreme cases ice can be thrown from blades up to several hundred metres.

If a risk from falling ice is considered to be significant and likely to cause damage to structures and vehicles or injury to the general public an 'ice safety zone' can be designated. Although the occurrence of a blade throwing a chunk of ice of a size great enough to pose a genuine threat to safety will be low (as ice on blades destroys the 'lift force' generated) the following formula can be used for calculating a safety zone around a turbine: 1.5 x (hub height + rotor diameter).[9]

If ice fall is found to be a persistent issue after installation, the turbine could be prevented from turning when the temperature goes below a certain temperature threshold.

Clearly, individual and group risks for all possible accident scenarios would have to be quantified for a major urban wind energy development as part of the risk analysis contained in the overall environmental impact assessment (EIA). Risk analyses are also, of course, carried out for conventional wind farms, but urban installation may present further complexities. For example, the risk of children attempting to climb towers or throwing objects at the blades.

It should be noted that there are many examples of turbines which allow the use of the ground beneath the blades of a turbine by the public (or animals in the case of turbines located on farm land).

Visual effects

For traditional onshore wind farms the visual aspect of the development is usually something to minimize. This mindset can be extrapolated to the urban wind energy situation, i.e. where the technology is seen by some as something that should be hidden away. One example has seen planners limiting the tower height on a 5m diameter urban turbine sited next to an eco-centre which is designed to promote and facilitate sustainable

living. Although the device is quite fitting and could be seen as a symbol drawing attention to the centre, the planners' view was to minimize the visual impacts. This comes at the cost of not only reducing the potential positive visual impacts but also significantly reducing energy yields.

Visual effects of larger structures in urban areas usually relate to the impairment of nationally or locally designated buildings, monuments or areas of importance to the landscape in the vicinity. Wind turbines should also not be incongruous or overly dominant components of the local or distant views. In small towns a reasonable sized turbine could serve to create a focal point, or dynamic monument, which could become of importance to the townscape.

An established methodology for assessing visual impacts[10] would include:

- a desktop study of the existing landscape character for the catchment area;
- zone of visual influence (ZVI) studies to identify key viewpoints which could be affected;
- identification of key groups affected;
- photomontage construction from agreed key viewpoints before and after the completion of the proposed wind energy development;
- an assessment of the significance of the effect on the landscape character.

The aesthetic quality of a turbine can be important especially if it is intended to create a landmark for a local area. For larger scales, three-bladed turbines are most common although two-bladed turbines are available. These turbines range from high aesthetic quality to lower aesthetic quality. Generally, three bladed turbines are considered more aesthetically appealing. Figures 3.10 and 3.11 attempt to demonstrate this distinction between what may be perceived as high and lower aesthetic quality in turbine design.

The design of the Enercon turbine has been aided by Foster & Partners and the E-66 turbine has been honoured with 'Millennium Product' status by the Design Council (UK).

Figure 3.9 Can turbines be appropriate to the urban landscape? Green Park turbine, UK (Ioannis Rizos)

Figure 3.10 Example of a what could be considered a lower aesthetic quality turbine

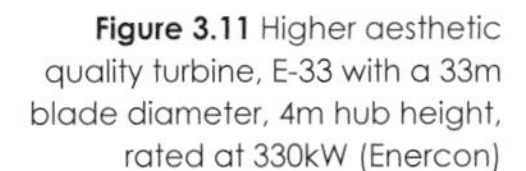

Figure 3.11 Higher aesthetic quality turbine, E-33 with a 33m blade diameter, 4m hub height, rated at 330kW (Enercon)

Noise

Securing planning permission for most wind energy-related developments requires evidence that applicable noise regulations will be satisfied during both daytime and night-time.

Most European countries (e.g. the UK, Germany, The Netherlands) have statutory legislation to regulate general noise level limits and, in many cases, specific guidelines and recommendations setting out advice for the assessment and measurement of noise from wind farms. These generally will relate to rural areas where the background noise is very low (< 40dBA).

In contrast, in urban locations, the ordinary background noise levels can reach 70dBA. The lack of precedents for the siting of wind turbines in urban/residential locations will usually mean that planning conditions are set on a case-by-case basis based on the existing noise regulation relative to the urban environment.

Two types of noise are associated with wind turbines:

- **Aerodynamic**, where the noise is radiated from the blades and is mainly associated with the interaction of turbulence with the surface of the blades. The turbulence may originate either from the natural atmospheric turbulence present in the wind or from local viscous flow in the boundary layer around the blades.

- **Mechanical**, normally associated with the gearbox, the generator and the control equipment, and perceived in general as an audible tone which is more intrusive than a broadband noise of the same sound pressure level. The noise is transmitted along the structure of the turbine and radiated from surfaces, for example, the nacelle-raft, the towers and the blades.

The predominant mechanism of noise generation from larger wind turbines is now the aerodynamic noise radiated from the blades, as insulation of the turbine nacelle and isolation of machinery parts using well-established vibration control techniques can reduce mechanical noise by up to 15dBA, and a number of larger machines ('direct-drive' models) have now dispensed with gearboxes altogether. For small machines, the noise of the turbine is often imperceptible due to the general noise of the wind, particularly at moderate wind speeds.

'Noise emission' can be defined as the sound power level (SWL) created by noise sources. 'Noise propagation' refers to the way that noise emitted by the wind turbine propagates in the local environment to the observer who then perceives a sound pressure level (SPL).

Sound spreads out from a point source as an expanding spherical 'surface'. Consequently, the sound energy is spread over an increasing area as it travels further away from the source. Therefore, taking only distance into account, sound levels are known to decrease by about 6dBA as the distance from the source doubles.

A 5dBA sound level change would probably be perceived by most people under normal listening conditions, although it would take ideal listening conditions to detect sound level differences of 2 or 3dBA. But care needs to be taken when evaluating sound as individuals can perceive a 10dBA increase in a noise source as a doubling of loudness.

Additional factors to be considered, apart from distance and the noise duration, include the frequency of the sound, the absorbency of the intervening terrain and the presence or absence of obstructions. The topography and presence of structural barriers such as walls may absorb, reflect, or scatter sound waves and can mitigate or increase noise levels. Wind speed and direction, humidity levels and temperatures can also affect the degree to which sound is attenuated over distance. Wind speed gradients ('wind shear') can also bend sound over distances of say >100m. In some cases, depending on the prevailing wind direction, the 'wind shear' may either bend sound upwards and away from certain sensitive areas (e.g. residential areas) or down towards buildings. The wind shear may also affect the character of the noise and the turbulence level will also affect the noise levels. Therefore, noise measurements taken from a particular turbine in one location may be different to those recorded if the same turbine was in another location (with a different 'wind shear' and turbulent character).

The impact also depends on the ambient sound levels, the time of day and who is perceiving the sound. Noise, like the thermal comfort of a building, is a subjective issue (i.e. one person may find certain conditions more or less acceptable than another individual).

Table 3.2 Weighted sound levels and human response

Source/activity	Indicative noise level dBA	Human response
	140	Threshold of pain
Jet aircraft at 250m	105	
Shout (15cm)	100	Very annoying
Pneumatic drill at 7m	95	
Heavy truck (15m)	90	Hearing damage*
Motorway traffic at 15m	70	Intrusive
Truck at 30mph at 100m	65	
Busy general office	60	
Car at 40mph at 100m	55	
Normal speech at 5m	50	Quiet
Wind farm at 350m	35–45	
Soft whisper at 5m	30	Very quiet
Rural night-time background	20–40	
Broadcasting studio	20	
	10	Just audible

*(8-hour exposure)

Good public relations should not be dismissed in being a real benefit in mitigating borderline cases. It has been said that someone will have a problem with noise if they want to have a problem. The general human response to various sound levels (together with comparative sources) are given in Table 3.2.

Fortunately sound is measurable and so direct evidence can be given in a planning submission to address this subject with a reasonable amount of confidence. Noise emission and propagation from a turbine can also be simulated using specialist software which takes into account the relevant factors such as the presence of the building, trees and roads (tarmac and concrete are very acoustically reflective).

Noise emissions measured from a 7m diameter, 10kW turbine, as part of research conducted by the National Renewable Energy Laboratory (NREL), are given in Figure 3.12.[11] The measurements were taken a slant distance (i.e. from rotor to microphone) of 54m and show how at this distance the noise associated with the turbine operation can be almost imperceptible from background noise in a rural environment over a variety of wind speeds. Urban background noise, even at low wind speeds, is often much higher that the background noise data in this diagram.

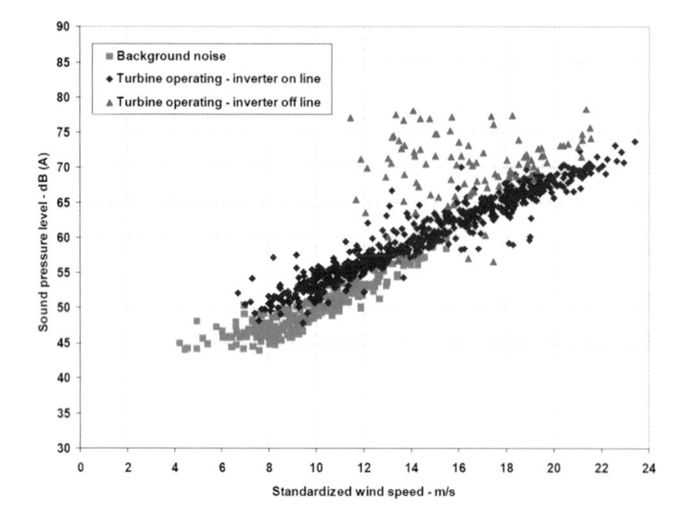

Figure 3.12 Example of noise emissions data (for a rural site) for a 7m diameter, 10kW turbine, a slant distance of 54m from the microphone (National Renewable Energy Laboratory, Colorado, USA)

A turbine produces much more aerodynamic noise when the inverter is manually switched off-line, i.e. when no electricity is being generated and the turbine blades are allowed to rotate freely. It should be noted that when the inverter is on-line, and electricity is being generated, a resistance is produced between the rotor and the stator which prevents the rotor from spinning uncontrollably.

Different wind turbines (and towers) produce markedly different qualities and levels of sound. In the past noise has been an issue with certain wind farms and the cumulative effect of having many turbines can exacerbate problems. Fortunately, technological improvement and increased understanding of the mechanisms associated with noise generation has ushered in new manufacturing processes and quieter turbines. Manufacturers are very tuned to this subject and there are several techniques that can be employed to mitigate noise issues (the most common being, as mentioned, gearless variable speed turbines, and also improved blade design and lower tip speed ratios).

The best way to appreciate the noise of larger gearless variable speed turbines is of course to visit them under varying weather conditions.

Smaller turbines generally produce less noise and in some cases are almost completely silent. For example, the Windside VAWT turbine (shown in Figure 4.1, page 118) claims to have practically

zero sound emissions as a consequence of using 'drag forces' to drive the shaft (i.e. slower rotation than 'lift force' turbines). The Quiet Revolution website[12] provides an example of the quality of the noise produced by their turbine (although a site visit or direct measurement would be necessary to gauge the amplitude at various distances from the device).

Often during a feasibility study for larger projects it may not be possible to say categorically whether a specific turbine will produce noise problems. This matter can be resolved after a number of key decisions are made, using the results of the study, such as which specific turbine to select (each will have their own emission character) and which location is preferable (including tower height).

At this point, if there is a likely problem, an acoustic survey can be carried out. This acoustic survey will allow:

- ambient noise levels of all audible frequencies to be monitored during different times of the day at key pedestrian points;
- these ambient noise levels to be compared to noise emissions information as provided by the manufacturer of a potentially suitable turbine.

Ambient noise levels will decrease at night as there will be less noise, e.g. from traffic, and so sound originating from the turbine would become more audible during this period. Should mitigation be required, e.g. if night-time levels are too high for the nearest homes, one of the most reliable techniques for improving conditions is simply to site the turbine further away.

Should noise problems occur, some turbines may allow full blade pitch control (e.g. large-scale turbines) and so the blades could be programmed to partially pitch out of the wind during certain conditions (e.g. at certain times and wind speeds) to reduce the rotation rate and associated noise levels. Other smaller/simpler turbines could have other automatic speed controls applied at appropriate times, although this is far from an ideal condition and should be viewed as an insurance policy on a system that is expect to function without problem.

Generally, close proximity to residential buildings will mean noise considerations are given a very high priority. Guidelines and criteria for continuous noise emissions are available from several sources. Wind turbine noise recommendations from PAN 45 (Planning Advice Note Rev 2002) are based on 'The assessment and rating of noise from wind farms' (ETSU Energy Technology Support Unit for DTI 1996). This recommends that the standard approach to controlling wind farm noise is through the use of limits made at the closest noise-sensitive properties (or to external areas which are designated for activities where quiet is highly desirable). It should be noted that any recommended limits should take into account the simultaneous variation of turbine noise and background noise with wind speed as well as factoring in the cumulative effect of any other wind turbines in the area.

Separate noise limits are used for daytime and night-time and this helps to ensure sleep disturbance is accounted for in residential areas. The guidelines state:

> Noise from the wind farm should be limited to 5dBA above background for both day and night-time, remembering that the background level of each period may be different.
>
> If the noise is limited to an $LA_{90,10min}$ of 35dBA up to wind speeds of 10m/s at 10m height, then this condition alone would offer sufficient protection of amenity, and background noise surveys would be unnecessary.
>
> The $LA_{90,10min}$ descriptor should be used for both the background noise and the wind farm noise, and when setting limits it should be borne in mind that the $LA_{90,10min}$ of the wind farm is likely to be about 1.5–2.5dBA less than the LA_{eq} measured over the same period. The use of the $LA_{90,10min}$ descriptor for wind farm noise allows reliable measurements to be made without corruption from relatively loud, transitory noise events from other sources.

$LA_{90,10min}$ is the ten minute average of A-weighted sound power levels exceed for 90 per cent of the time (A-weighted is the filter which represents the ear's response to sounds power levels by desensitizing the lower frequencies).

Table 3.3 summarizes the limits for continuous wind turbine noise for night-time from ETSU[13], the World Health Organization[14] and British Standards[15].

	Energy Technology Support Unit (ETSU) 'The assessment and rating of noise from wind farms'	**World Health Organization (WHO) 'Guidelines for community noise'**	**British Standard BS 8233:1999 'Sound insulation and noise reduction for buildings'**
Bedroom	35dBA (sleep disturbance criteria)	30db LA_{eq} (sleep disturbance criteria)	30dB LA_{eq} (good) 35dB LA_{eq} (reasonable)
Associated night time fixed limit	43dBA*	45db LA_{eq}**	–

Table 3.3 Wind turbine continuous noise limits for night time

* using an allowance of 10dBA for attenuation through an open window (free field to internal) and 2dBA subtracted to account for the use of $LA_{90,10min}$ rather than $LA_{eq,10min}$

** using a 15dB reduction across an open window.

Interestingly, the ETSU guidelines state that in 'low noise environments' the day-time level of the $LA_{90,10min}$ of the wind farm noise should be limited to an absolute level within the range of 35–40dBA, with the actual value chosen by considering factors such as the number of dwellings and the effect of noise limits on the number of kWh generated. However, the daytime and night-time limits can be extended to 45dBA when the occupant has some financial interest in the wind energy development.

Shadow flicker and blade-reflected light

Shadow flicker is caused when the sun passes behind the spinning blades of a wind turbine. The rotation of the turbine blades creates a shadow, which will flicker at different rates depending on the speed of the blades (which will vary depending on the model and wind conditions). This phenomenon can affect people in neighbouring residential and office properties and may have the greatest effect on those suffering from epilepsy. Around 0.5 per cent of the population are epileptic and 5 per cent of the population are photosensitive. Of the photosensitive proportion less than 5 per cent are sensitive to low frequencies (2.5–3Hz) with the remainder sensitive to higher frequencies. Flicker problems are rare and in general the problem occurs in buildings where the flicker appears through a narrow window opening.

Table 3.4 provides an approximate flicker sensitivity guideline (with a corresponding turbine rpm rating) illustrating how the degree of nuisance depends on the rate of flickering. In reality the actual critical flicker frequency increases as the light intensity increases up to a maximum value, after which it starts to decrease.

Flicker rate (Hertz)	Human perception	Equivalent RPM for 3-bladed turbine
< 2.5	Negligible effect	<50
2.5–3	May affect 0.25% of population	50–60
3–10	Effect is perceivable	<200
10–25	Greatest sensitivity	200–500
> 50	Continuous light source	1000

Table 3.4 Flicker rate and human response

The operational speeds of smaller turbines are often below 150rpm (such as Proven WT6000 and WES5, shown in Figures 2.4 and 2.6 on pages 41–42). The operational speeds of larger turbines are lower still, e.g. Enercon E33 (33m diameter blade) which operates typically between 18 and 45rpm.

If a problem is anticipated with a particular turbine the difficulty may be in determining the extent of the area over which the flickering effects need to be analysed. The extent could potentially be quite wide as winter shadows cast for some distance due to the lower solar altitude. However, the flickering effects have been proven to occur up to ten rotor diameters from a turbine.[16] Therefore, a flicker map can be created by overlaying concentric rings around a turbine location marked on a scaled site map. Ring intervals of 100m will correspond approximately to turbine blades with diameters of 10m, 20m and 30m etc. However, flicker effects will not occur in the whole of the concentric zone. In countries north of the equator there will be a region to the south of the turbine where no flickering can occur. For example, in the UK, flicker problems cannot occur approximately 50° from either side of due south of the turbine. Similarly, in countries south of the equator there will be a region to the north of the turbine that cannot be affected.

When assessing flicker the direction of the prevailing wind should be taken into consideration in order to determine whether any 'sensitive areas' align with the full face of the turbine or a narrow profile.

Other factors should also be considered. For example, longer winter shadows may be mitigated by the fact that during this period, when the sun is low, the flicker is more likely to be obscured by either cloud on the horizon or intervening buildings and vegetation. Also, around midday the flickering should not extend far (especially in summer, when the sun/flickering is strongest, due to the high solar altitudes).

In many cases, any nuisance that does occur from flickering is likely to be brief and occur for only a short period every year if, that is, at that moment the blades are rotating at the correct frequency (they do not always rotate), the sky is clear and the area is occupied. As a final mitigation strategy, should any complaints of flickering arise, the turbine can be stopped (or slowed) at certain times of the year.

The area blade flicker may effect is quantifiable. Should planning authorities wish to see evidence that blade flicker will not be a problem, computational 'ray tracing' shadow sequences can be generated. These use an accurate 3D model of the particular area and can identify particular windows of neighbouring properties which may encounter this phenomenon as well as estimate the number of days this would occur and duration of such events.

Flickering on the roads may bring about additional concerns as this could conceivably bring about a distraction for drivers. Precedents do exist, as mentioned, where large turbines are sited near roads and these cases should be examined in order to assess the potential risks, e.g. the Green Park Turbine sited next to the M4. The UK planning guidance[17] states that:

> Drivers are faced with a number of varied and competing distractions during any normal journey, including advertising hoardings, which are deliberately designed to attract attention. At all times drivers are required to take reasonable care to ensure their own and others' safety. Wind turbines should therefore not be treated any differently from other distractions a driver must face and should not be considered particularly hazardous. There are now a large number of wind farms adjoining or close to road networks and there has been no history of accidents at any of them.

Apart from shadow flicker, problems can be caused by the suns light reflecting off the blades which can produce a flashing effect visible for some distance. This can be mitigated by choice of finishing. Light grey semi-matt finishes are often used, however, other colours and patterns can also reduce the effect further. Additional information can be found in 'The influence of colour on the aesthetics of wind turbine generators' – ETSU W/14/00533/00/00.

Electromagnetic interference (EMI)

Any large obstacle, such as buildings or wind turbines, can cause interference over a wide frequency range to nearby radio and television systems, microwave links and satellite services as well as to the radar, guiding and landing systems of/for aircraft. As such there are a number of statutory consultees for wind energy proposals. These usually relate to wind farms; however, the scale of a given urban wind energy project may be large enough to warrant communications, at the early stages, with the following (who have objected to larger turbines and wind farms in the past): ministries concerned with national defence and radar, telecommunication companies, aviation authorities and national air traffic services. In some countries, such as the UK, airport radar regulations can be strictly enforced for any turbine within 30km of an airport and therefore limit the permissible size of turbine that can be erected. The British Civil Aviation Authority provides the following comments on the impact of wind turbines:

> A wind turbine within direct line of sight of a radar station can create a radar echo strong enough to swamp the radar receiver, thereby causing a large point to appear on the screen which masks aircraft echoes.
>
> A turbine disc can also cause a radar shadow, preventing visibility of targets beyond it on the same line of sight. Beams reflected from blades and terrain features can cause aircraft to be reported at incorrect bearings.
>
> An aircraft flying behind a turbine and partly obscured by it can jitter on the radar screen, or appear in a location skewed from its real one, due to beam diffraction.
>
> Turning down the sensitivity of a radar to remove extra clutter caused by windfarms, a radar operator can lose returns from poorly reflecting targets such as small aircraft.

However, many countries simply follow Civil Aviation Organization guidelines which may seek to limit the height of any tall structures within the airport catchment area. The mechanisms creating electromagnetic interference and

Figure 3.13 Wildlife and turbines (Ioannis Rizos)

resonance from conventional HAWTs are complex to predict. High-frequency interference depends on small structural details, while low-frequency interference depends on broad characteristics such as the height of the turbine tower and the number and rotational speed of the blades.

For turbines integrated into the built environment, proximity to a telecommunications system may necessitate investigation. Mitigating factors include:

- Most turbine blades are made of glass-reinforced plastics (GRP), i.e. they are non-metallic.
- The turbines integrated into a built environment are relatively small in scale compared to those used in conventional wind farms.
- Wind farms often have many turbines in rows (on hilltops) and this can cause interference – one turbine in isolation will tend to have a much lower impact.

As the impact on local television reception is difficult to predict an applicant may have to support the inclusion of a planning condition requiring mitigation should significant TV reception interference occur after the installation of a large turbine. However, guidance and tools to help assess the impact do exist.[18]

Biodiversity and birds

In the past this subject has created some heated debate between the wind industry and campaigners. Mistakes have been made where turbines have been sited in migration paths; however, consultation with avian experts is now a standard part of wind farm development in order to avoid such misfortunes.

The majority of wind farm owners treat the issue with great seriousness and more facts are emerging through monitoring and the running of various trials. For example, research is being carried out with the aim of producing turbine designs which minimize bird kills, e.g. by changing the colour and pattern on the blades or reducing the 'perching' opportunities of a turbine. However, correlations are difficult due to the rarity of occurrence. Some research indicates that, generally, around one to three bird kills per year occur for each large-scale turbine (perhaps as certain local bird populations become aware of these features and adapt accordingly).[19]

Nevertheless birds and other flying creatures such as bats are killed by turbines. However, it should be kept in mind that bird kills through turbine strikes are dwarfed by statistics related to domestic cats, building strikes and impacts with vehicles.

On a wider scale, it is clear that the continued use of conventionally fuelled power plants could eliminate many hundreds or even thousands of creatures as well as entire species through, for example, climate change, acid rain, and pollution. The continued emissions of pollutants such as greenhouse gases, sulphur dioxide, mercury, particulates, or any other type of air pollution is also accompanied by the other forms of damage to the planet, e.g. via the extracting and transporting of resources or in the event of accidents.

Siting wind energy near the end users and away from natural habitats is usually more favourable in terms of limiting the impact on birds and bats. If a particular urban wind turbine is considered to be a threat to wildlife, the planning authority (after consultation with local wildlife organizations) can stipulate in a planning agreement that bird kills are monitored. Bird specialists can be consulted, especially if there are rare species such as birds of prey in a particular area.

Property values/house prices

Should the local environmental impacts be poorly considered, the effect on local property values, especially for residential buildings, may be considerable. There are several cases where onshore wind farms have been built close to homes and noise (or visual penetration, including flicker) has created an undesirable environment. The stress from problems such as interrupted sleep are compounded by the stress caused by knowing that others will not be interested in living under the same conditions and that the property is correspondingly devalued. A sense of being 'trapped' in these conditions would undoubtedly heighten any sense of intrusion. As stated previously, a given development should not benefit the global environment to the detriment of the local environment. In some of these cases, the owners have been appropriately vocal about the issue.

Although theses cases constitute a very small percentage of those in the vicinity of wind turbines, alarm can be extended to others who may feel they may be exposed to a similar situation should a proposed wind development go ahead. It is difficult for a lay person to know the effects of any wind energy proposal and only experts in the field can begin to fully judge the impacts. Several studies have been carried out to attempt to gauge the effect of turbines on house prices. However, this is not an easy subject to resolve with any great degree of certainty as house price fluctuations are a result of many complex factors.

A report entitled 'What is the effect of wind farms on house prices?' has been produced by the Department of Real Estate and Construction, Oxford Brookes University, UK[20] with support from a grant from The Royal Institution of Chartered Surveyors (RICS). This report examines the impact of wind farm developments near residential properties in Cornwall, UK. It was concluded that although initial evidence pointed towards an effect of turbines on house prices, on closer inspection it appeared that 'there were generally other factors which were more significant than the presence of a windfarm'. Studies from the British Wind Energy Association (BWEA) and the Renewable Energy Policy Project (REPP) in the US were also summarized. The former found that there is a 'detrimental effect on [property] values either due to close proximity of the wind farm or its visibility'. The latter, after examining 24,300 property transactions, concluded there was no evidence to suggest that wind turbines sited within a five mile radius of a property had a negative impact on value. Furthermore, evidence to the contrary was found, i.e. wind turbines had a positive effect on property value.

Interestingly, Dent and Sims from Oxford Brookes state: 'There is evidence to suggest that the "threat" of a windfarm may have a more significant impact that the actual presence of one.'

The report also mentions the term 'NISEBY' which describes a phenomenon they encountered ('not in someone else's back yard') where the majority of planning permission objections originate from areas that are extremely far away from the proposed site. These types of actions may come from those who have been unduly affected by turbine development in their own area and who have concern for others. Alternatively this response may originate from those who feel they are against wind energy development as a whole. Objection to traditional wind farms may stem from the belief that wind energy developers are erecting turbines to make money without any regard for the impact on others – such as the effects on health, property value, animals/birds, the intrinsic beauty of the countryside and the livelihood of those who depend on local tourism. The idea that developers can make

money from government incentives such as feed-in tariffs and ROCs, where the taxpayer or electricity customer foots the bill, may be difficult to swallow especially if accompanied by the belief that these turbines have no significant associated global benefit.

This issue of property devaluation may be of more importance than given on first inspection. Although it may not be directly cited as a reason to object to a wind energy development, it may be the number one reason behind many objections.

Transparency, involvement and, in particular, co-ownership from the community are a means for wind energy developers (including urban wind energy developers) to allay fears and gain local support. This could involve the presence of a representative at the disposal of the community and even representing their interest. The value of providing a means for the local community to express their concerns or support should not be underestimated. Sims and Dent state: '[Wind energy] may not translate into lower house prices if the community are actively involved in the process and enjoy some of the benefits through lower, or greener, fuel costs.'

4) ECONOMIC ASPECTS

Given the importance of the financial aspects of wind energy developments, the economic viability has been commented on in Part 1 and crudely demonstrated for a large-scale stand-alone turbine. As mentioned, payback periods can vary hugely and there are cases, especially where wind resources are unexpectedly poor, where this period will extend past the design life of the turbine.

Aside from cases where energy generation is lower than anticipated, there will be situations where additional costs are discovered as a project progresses. Increases in capital costs will of course also increase the payback period. Accurate costing often requires experienced personnel to identify and quantify sources of expenditure. Some of these sources are discussed in this section.

Wind energy development may be economically evaluated in several ways. Simple predicted payback, as described in this section, is probably the most common method used for urban wind energy, although some projects are evaluated using return on investment (ROI) models. Urban wind energy developments will tend to have a low ROI unless the energy price increases are well above inflation. Although most individuals would prefer energy prices not to follow on from the recent steep increases, this model may not be unrealistic (at least in the short to medium term as shown by oil price trends).

As already suggested, the pure ROI viewpoint can be thought of as relatively short-sighted in terms of wider social and environmental needs, and investments can be made that take into account individual wealth as part of community, regional, national and global wealth. This idea is exemplified by the wind turbine in Toronto shown in Figure 3.14. In 2002 the fixed-speed Lagerwey, which stands almost 100m high to the tip of the blade, was erected in a prominent location by a 400-strong 'Windshare Co-op' co-ownership scheme.

Figure 3.14 Toronto turbine manufactured by Lagerwey (now Emergya Wind Technologies manufacturing mid-range turbines available with either 750kW or 900kW) (Harold L. Potts)

Simple predicted payback assessment

At the initial feasibility stage of a project there are limits to the level of economic detail that can be examined. However, a preliminary assessment can provide useful guidelines and help inform the decision-making process. The most common method to express the economic value of a project – 'predicted payback' – is considered in two main categories: costs and returns. These categories can be further split into 'initial' and 'running' items.

Initial costs include equipment costs, installation costs (including delivery and commissioning), foundations, access or 'hard standing', grid connection charges, crane hire and design fees. The running costs include operation and maintenance (O&M) costs, insurance costs (e.g. public indemnity insurance), possible land rental (which may take the form of a percentage of the cost per kWh) and contingencies relating to remedial work or compensation claims.

A summary of approximate initial costs for a range of turbines is given in Table 3.5. As may be expected given relatively large amount of investment, large stand-alone turbines represent more value per square metre of swept area over smaller wind turbines. However, although the cost of large-scale wind energy still continues to decrease, there is also considerable scope for price decreases in many smaller turbines if uptake increases and the markets are stimulated.

Table 3.5 Indicative equipment and installation costs

Category	Type and size	Swept area (m²)	Budget costs	Normalized £/m²
Large stand-alone	70m diameter (2MW) HAWT	3850	£1,000,000	1.0
Large stand-alone	50m diameter (0.7MW) HAWT	1257	£400,000	1.2
Small stand-alone	9m diameter HAWT	64	£50,000	3.0
Small stand-alone	5m by 3m VAWT	15	£50,000	12.8
Small stand-alone	5m diameter HAWT	20	£25,000	4.9
Micro 'home' turbines	2m diameter HAWT	3	£3000	3.7

Although refurbished turbines will often be a cheaper option in terms of initial costs, their design life will be less and so this option may in some cases be a false economy.

Running costs are usually very low as turbine operation is fully automatic with turbines self-starting (at the cut-in wind speed) and stopping when speeds are either too low or too high (via an automatic cut-out mechanism to prevent equipment damage). Maintenance usually requires low-level infrequent checks on meter readings and reporting on any problems such as signs of damage, checking for evidence of bird kills, or noting any change in noise emission. More involved maintenance checks may include adding lubrication or changing oil, adjustments to mechanical equipment and checking control equipment. In some cases these more involved procedures may not take place for several years at a time.

The Energy Saving Trust[21] calculated the likely initial costs and maintenance costs for large-scale wind turbines. These were around £350,000 for a 600kW turbine (50m blades) and a total of 5.5 per cent for maintenance costs.

The initial returns will arise from any grant payment released on completion of the project. The running returns arise from electricity sales or savings or from such schemes as the Renewable Obligation Certificates (ROCs) in the UK (Box 3.4).

BOX 3.4

GOVERNMENTAL SUPPORT FOR RENEWABLE ENERGY PRODUCTION: RENEWABLE OBLIGATION CERTIFICATES (ROCs)

Since April 2002, UK energy providers are obligated to provide (until 2027) a certain percentage of their energy from renewable sources. The energy companies can achieve these targets by generating their own renewable energy or by paying others to do this for them. This is not to say the companies are 'buying the electricity' – just the fulfilment of an obligation. Thus energy companies could buy ROCs from any wind turbine in the UK whether the energy is consumed on-site or is sold to the grid.

The value of theses certificates varies and the Non-Fossil Purchasing Agency (NFPA) and e-ROC auctions have been selling each certificate for over £40 per MWh.[22] There will be a small cost (e.g. 50p/ROC) to the owner of a wind turbine in order to redeem their certificates via a broker (e.g. e-ROC) with a minimum charge (e.g. £300).

Once budget figures for the costs and returns are gathered, an approximate payback calculation is fairly straightforward to carry out. The confidence in this accuracy of the economic assessment will be strongly related to the confidence in the initial annual energy yield predictions. If this is high and further confidence in the economical assessment is required a more detailed calculation can be carried out.

Detailed methodology for assessing costs

One method that can be adopted is to calculate the effective price of a unit of electricity produced by integrated turbines based on an analysis over their expected operating lifetime. This method is more appropriate for developing national or utility energy planning policies.

However, the methodology described below is based on the widely used concept of 'net present value' (NPV). This is more appropriate from the perspective of a developer and extends the use of simple payback by incorporating real interest rates and inflation. NPV allows a cash flow, C, in N years time to be assigned a value (discounted) in today's terms by taking into account the cumulative impact of inflation in reducing the value of the future cash flow in real terms; and the opportunity cost, i.e. the fact that the developer could invest their money elsewhere and will require an acceptable rate of return on their investment within a given period of time.

NPV equation

The following equation can be used for an initial economic assessment of the potential for urban wind energy cases where one or more stand-alone machines are supplying neighbouring buildings and where the wind turbines are fully integrated into the building itself:

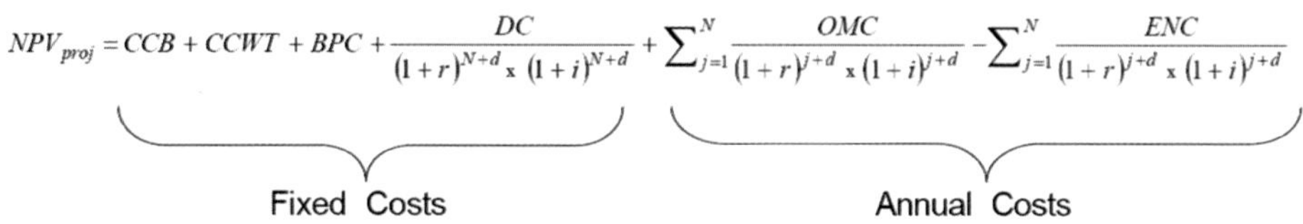

$$NPV_{proj} = CCB + CCWT + BPC + \frac{DC}{(1+r)^{N+d} \times (1+i)^{N+d}} + \sum_{j=1}^{N} \frac{OMC}{(1+r)^{j+d} \times (1+i)^{j+d}} - \sum_{j=1}^{N} \frac{ENC}{(1+r)^{j+d} \times (1+i)^{j+d}}$$

Where:

NPVproj	the net present value of the project
CCB	the marginal increase in the capital costs of the building from integrating the wind turbine(s);
CCWT	the capital costs of the wind turbine(s);
BPC	the balance of plant costs (normally expressed as a fraction of CCWT);
DC	the costs of decommissioning the turbine(s) at the end of their operating life(s);
OMC	the annual operating and maintenance costs;
ENC	the annual saving from displacing the need for grid-supplied electricity in the buildings and/or annual receipts from electricity exported to the grid;
r	the rate of return on the investment (test discount rate);
i	the rate of inflation;
d	the delay in years between the purchase of the wind turbine(s) and the beginning of electricity generation;
N	the amortization period or expected operating lifetime of the turbines (whichever is the shorter).

There are two ways of using the above equation:

- calculating the value of NPV_{proj} at either the end of the expected operating life of the wind turbine(s) (i.e. ~20–25 years) or after a shorter time specified by the developer (amortization period);
- calculating the 'break-even point' (number of years) at which $NPV_{proj} = 0$, indicating that the investment has been fully repaid and will generate a profit.

The developer will usually specify a so-called 'amortization period' defining the timeframe over which they expect to see a return on their investment, which may or may not be less than the expected operating lifetime of the wind turbine(s). Obviously, NPV_{proj} may still be positive when evaluated or a break-even point may not be achieved, in which case the developer will be subsidizing the use of the wind turbine(s).

For instance, they could choose to exclude the additional capital costs for the building, if the wind turbines were considered fundamental to the entire architectural concept. For reasons of simplicity, there are a number of implicit assumptions contained in the NPV equation:

- Both inflation and the desired rate of return on the investment (test discount rate) are assumed to be constant over the expected operating lifetime of the wind turbine.
- It is assumed that the investor has sufficient capital to begin with (rather than taking out a loan).
- All capital costs are assumed to occur at the same time. In practice a building can take several years to construct, and so, for instance, architectural and engineering design costs and costs of obtaining planning approval may be incurred well before electricity generation begins. The time delay is intended to account for the time period from the purchase of the wind turbine(s) until export (or displacement) of electricity starts (following installation and commissioning).
- All annual costs (i.e. money paid out for operation and maintenance or received for export of electricity to the grid) are assumed to be identical single payments at the end of each calendar year. Note that this assumes that these costs will fall in real terms (since they remain fixed despite inflation), which seems the likely scenario for electricity costs. Over 20 years, assuming a constant

2 per cent inflation rate, the unit electricity prices would effectively come down by one third. The energy actually produced by the turbine(s) can vary by as much as 10 per cent year on year depending on wind conditions.
- Energy costs are assumed to be negative costs (i.e. monies received).
- The equation is written in such a way that the costs for decommissioning at the end of the operating lifetime of the turbine are recognized as a fixed cost from the outset. This ensures that revenue from sales (or displacement) of electricity must generate an appropriate sum for payment of future decommissioning costs before the investment becomes profitable.

It is quite easy to see how further factors could be built into the equation for a more sophisticated cost assessment at a later design stage as more information becomes available. Multipliers could be introduced into the numerators to allow, for instance, electricity prices to rise at rates below, above or equal to general inflation. Similarly, if the lifespan of a building that housed an integrated turbine were to be, say, 50, 75 or 100 years (i.e. a major public building), it would be possible to factor in the cost of re-engineering the turbine(s) into the analysis and look at the overall calculation on a longer timescale.

Grants and fund-raising

If the scale of initial costs for an urban wind turbine project are found to be a barrier to progression there are a variety of opportunities to raise the necessary capital that can be considered. The Energy Saving Trust (EST) has published 'Thinking out of the box: Novel sources of funding for sustainable energy projects'.[23] It suggests a number of routes which could be adopted. One method that can be considered in some urban areas is to canvas local commercial entities within the catchment area of the development for 'sponsorship' (which may involve some form of advertising). The EST suggests typical reasons for companies wishing to give their support:

- to create goodwill and be seen as good citizens in the local community and more widely as a caring company;
- to associate themselves with a certain cause;
- peer pressure – because it is expected of them and their competitors are supporting projects;
- because a senior figure (e.g. chairman or chief executive) is interested in the cause;
- donating to good causes is tax-free.

Community ownership or share schemes can be initiated which help the local community to embrace a wind energy development. The Baywind Energy Co-operative, for example, now has 1100 members, 600 of whom live in Cumbria, UK, where Baywind part-owns the Harlock Hill wind farm. Baywind raised £2 million through share offers to its members. Voting rights are equally distributed irrespective of number of shares held, meaning that no individual or organization can have a controlling interest. Shareholders received a 20 per cent tax refund on their initial investment under the government's Enterprise Investment Scheme.

The Renewable Energy Investment Club (REIC) is an example of a not-for-profit organization that links renewable energy providers and ethical investors and acts as the facilitator for share offers in community developments. REIC currently has over 300 members with an investment potential of over £1 million. Through REIC, Bro Dyfi Community Renewables raised £54,000 for a wind turbine that supplies 45 households in the local area, with 94 per cent of shareholders from the Dyfi valley. This project also received £35,000 grant funding from the EST's Innovation Programme.[24] Third party finance schemes are now emerging to facilitate small wind energy developments such as power generator Ecotricity's Merchant Wind Power (MWP) initiatives for the turbine in the Sainsbury's Distribution Centre in East Kilbride. They can take on not just the process of financing, but also project management aspects, obtaining planning approval and construction of the wind turbine installation.

Governmental grants may also be available. For example, in the UK the Low Carbon Building Programme (LCBP) wind turbine grant application has presented opportunities to claim up to 50 per cent of the installed cost. Funds are distributed on a 'first come first served' basis until they are exhausted. To qualify for a grant there will be certain conditions that have to be met, e.g. to comply with the grant scheme basic energy efficiency measures must be installed in buildings supplied by turbines (therefore addressing holistically both energy demand and supply). These conditions depend on the grant 'stream' entered.

Importantly, grant application may directly affect decision-making. For example, grants may influence the project timeline or, for example, be available for only certain wind technologies (i.e. for specific products from a list of recommended manufacturers).

SUMMARY

When a prominent urban turbine is proposed, the impacts of such a device are immediately raised, relating primarily to environmental and economic aspects, together with issues pertaining to equipment siting and selection, and to the effective exploitation of opportunities for grants and funding, and positive public perception. Therefore an initial feasibility study is often considered a necessary step, especially when integrating a turbine into the built environment as this creates unique issues which have to be addressed beyond those of a standard wind turbine installation (e.g. on a conventional wind farm).

This feasibility study should provide a basis for a well-informed decision-making process. It will allow the developer(s) to assess where a proposal for a particular turbine stands between the potential benefits (such as capital and environmental gains) and the likely costs and local environmental implications. The feasibility study then can form the basis of both the planning application and a possible funding or grant application.

This study should touch on all of the relevant subjects and highlight the most important issues. Some subjects and issues addressed in these studies may only be discussed briefly to allow a scheme to be set in context. Their resolution will follow

Figure 3.15 Coastal turbines, Solway Firth, Cumbria, UK (Harold L. Potts)

from key decisions by the developer(s) and feedback from stakeholders (which includes the planners).

In many countries, an important aspect to keep in mind is public perception. Ensuring a positive public perception can be not only important to the success of the scheme but also for other important issues such as raising funds and obtaining planning permission. Failure to demonstrate community involvement, or opposition from members of the local community, can cause delays to permission being granted (or rejection) which will have cost implications. Some opposition can be anticipated from, for example, homeowners sited near a proposed site. These objections may primarily arise from concerns related to a potential decrease in the value of their properties. If deemed appropriate, public relations should be carried out in the early stages to present the facts and benefits of the development.

Other factors that can be considered to help ensure a positive public perception include turbine design (aesthetics), appropriate scale of the device, good performance (i.e. that the turbine will be seen to be turning for extended periods), good reliability (i.e. that the turbine is not in a state of disrepair for extended periods) and other issues such as low environmental impacts (e.g. appropriate noise emission characteristics).

REFERENCES

1 *Supplementary Planning Guidance (SPG) Wind Energy*, Newark and Sherwood District Council (July 1999) p3

2 T. R. Oke, *Boundary Layer Climates*, Routledge (1987)

3 www.berr.gov.uk/energy/sources/renewables/explained/wind/windspeed-database/page27326.html (September 2008)

4 Warwick Wind Trials Interim Report (May 2008)

5 www.nrel.gov/wind/international_wind_resources.html (September 2008)

6 www.wind-works.org/articles/DeathsDatabase.xls (October 2008)

7 Tammelin, Cavaliere, Holttinen, Hannele, Morgan, Seifert, and Säntti, *Wind Energy Production in Cold Climate* (WECO) (1997)

8 www.glsc.org/press/press.php?id=34 (October 2008)

9 Tammelin, Cavaliere, Holttinen, Hannele, Morgan, Seifert, and Säntti, *Wind Energy Production in Cold Climate* (WECO) (1997)

10 *Landscape Character Assessment Guidance for England and Scotland*, Scottish Natural Heritage and The Countryside Agency (2002)

11 P. Migiliore, J. van Dam and A. Huskey (National Renewable Energy Laboratory), *Acoustic Tests of Small Wind Turbines*, AIAA (2004)

12 www.quietrevolution.co.uk/projects.htm# (October 2008)

13 *The Assessment and Rating of Noise from Wind Farms*, Energy Technology Support Unit for Department of Trade and Industry DTI (ETSU-R-97)

14 *Guidelines for Community Noise*, World Health Organization, www.who.int/docstore/peh/noise/guidelines2.html (January 2009)

15 *Sound Insulation and Noise Reduction for Buildings*, British Standard BS 8233:1999

16 *Planning for Renewable Energy: A Companion Guide to PPS22*, ODPM (2004) p178

17 *Planning for Renewable Energy: A Companion Guide to PPS22*, ODPM (2004)

18 *The Impact of Large Buildings and Structures (including Wind Farms) on Terrestrial Television Reception*, BBC & Ofcom

19 *Wind Turbines and Birds*, AusWEA (Australian Wind Energy Association)

20 Peter Dent and Sally Sims, *What is the effect of wind farms on house prices?* Department of Real Estate and Construction, Oxford Brookes University, UK

21 www.est.co.uk/uploads/documents/housingbuildings/case5.pdf (October 2008)

22 www.nfpa.co.uk/ (July 2008)

23 *Thinking Out of the Box: Novel Sources of Funding for Sustainable Energy Projects*, Energy Saving Trust (EST) www.cse.org.uk/pdf/pub1032.pdf (October 2008)

24 www.reic.co.uk (November 2008)

Turbine Technology

'North America's first urban Wind Turbine' in Toronto, Canada (Harold L. Potts)

INTRODUCTION
DESIGN PRACTICALITIES

A project viewed favourably by relevant parties after a first-pass feasibility study can progress to the subsequent stage. This will require the design details to be fleshed out and a more in-depth knowledge of turbine technology. Therefore, this section serves to inform project leaders and designers on key technological issues and practical aspects. This includes elaboration of a number of terms alluded to during the preceding sections (such as drag and lift forces, coefficients of performance, tip speed ratio and overspeed control) through the illustration of the fundamental principles behind these key topics. Concise descriptions are provided to allow timely progression, avoiding the less easily digestible 'higher level' of knowledge required by manufacturers and equipment designers.

A focus on the practical side is maintained throughout, which entails highlighting the possibilities and pitfalls of working with the current available turbine technology.

This section covers the following:

1 Turbine types
2 Generator types
3 Blade design
4 Protection
5 Tower design
6 Grid connection
7 Sourcing equipment
8 Wind energy yield enhancements techniques

The information provided in this section leads on to the final section which covers additional design aspects related to building-integrated turbine technologies.

1) TURBINE TYPES

HAWT vs VAWT

The two types of turbines, horizontal axis and vertical axis wind turbines (HAWTs and VAWTs), have been mentioned throughout this text. For large-scale turbines, the market has converged on the three-bladed horizontal axis turbine as the 'right way forward' for multi-megawatt turbines. Furthermore, the main manufacturers now only produce these types and so there is little option for most to opt for large-scale VAWTs. For small-scale wind energy both types are readily available and a variety of designs are available in both categories. Wind energy projects can have wide variations in priorities and environmental considerations. A good understanding of fundamentals will allow the selection of the appropriate kit for the particular project.

For small scales, the VAWT could prove a significant contender to the historically favoured HAWT. Although long-term data on the performance of VAWTs are currently not widely available, VAWTs have (in theory) several advantages over horizontal wind axis turbines such as:

- less maintenance as there are fewer and slower moving parts (as they can have lower rotation speeds and there is no yawing mechanism to turn the blades into the wind);
- they emit less noise (as they can have lower tip speeds and air compression from a blade passing the tower is eliminated);
- turbulence and winds from all directions (not just horizontal winds) are handled more effectively, which can be an important factor when integrating turbines into the built environment;
- some designs can be perceived as more aesthetically pleasing.

However, horizontal axis turbines continue to dominate as they are cheaper (requiring less material per square metre of 'swept area') and have received the most attention/funding.

Although more material is required for a VAWT per square metre of 'swept area' there is still significant scope for the continuation of the downward trend in equipment costs if uptake increases. However, there are other issues to consider e.g. 'overspeed' control for Darrieus type VAWTs (i.e. how to avoid damage to the turbine in wind speeds >12m/s) as discussed later.

Lift vs drag

Both HAWT and VAWT turbine types can be split into sub-categories depending on whether they primarily make use of the so-called 'lift' force or the 'drag' force to turn the rotors (see Box 4.1).

For HAWTs, it is only traditional windmills (e.g. used to process grains or pump water) that are designed to primarily make use of the drag force to turn the blades. Simple drag-force turbines have their speed naturally limited to the maximum wind speed. Therefore these traditional windmills are characterized by slow-moving blades with large surface areas. All modern day HAWT turbines are designed to make use of the 'lift' phenomenon and, as such, are characterized by fast-moving blades with low surface areas.

BOX 4.1

THE ORIGIN OF THE 'LIFT' AND 'DRAG' FORCES ON TURBINE BLADES

The forces distributed over the surface of the blade resolve (combine) into the two components of lift and drag. This is illustrated by the example provided of a generic blade designed to promote lift force and minimize drag.

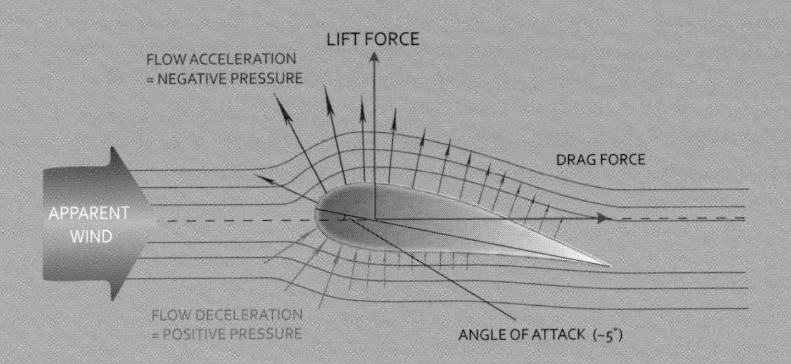

The drag force is the combined component acting parallel to the direction of the 'apparent' wind flow (see Box 4.2). The lift force is the combined component acting in a direction perpendicular ('normal') to the 'apparent' wind flow direction. A high performance blade will have a high lift-to-drag force ratio (e.g. >100).

For VAWTs, lift turbines again tend to dominate (Darrieus type). However, drag force turbines (Savonius type) are also available. The vast majority of lift-force turbines on the market significantly outperform the drag type turbines, i.e. they extract more energy per square metre of swept area. However, these drag type VAWTs do present potential for many situations as they can have the following properties:

- low risk (typical sturdy heavyweight construction can reduce the perception of risk of, for example, blade shedding and the lower rotation speed can reduce the impact on birds);
- highly reliable (coping with high wind speeds – e.g. up to 60 m/s – without cutting out, and capable of operating in extreme conditions such as the Antarctic or on oil rigs);
- low maintenance (for example, self-lubricating bearings and five-year maintenance cycles);
- quiet (due to the low rotation speeds);
- low start-up speed (so will be seen to be generating energy for longer periods);
- visually more attractive.

Figure 4.1 Windside (drag force type) vertical axis wind turbines, Synergia in Oulu, Finland

The relative low power production of drag type VAWTs stems from the low swept area available to capture the energy in the wind and the lower coefficient of performance. Although the performance can be relatively high at low wind speeds, the efficiency tails off during high speed (energy-rich) winds. For example the Windside WS-4A (4m high by 1m wide) has an overall Cp of ~30 per cent at 3m/s and 20 per cent at 6m/s. It produces 20W, 100W and 400W at 3m/s, 6m/s and 10m/s respectively.

These turbines can be considered as 'dynamic art'. An 'artistic' cluster of 4 vertical axis turbines (each rotating at different speeds) from the Finnish manufacturer Windside is depicted in Figure 4.1. The largest turbine in this figure is 4 x 1m although they produce a 6 x 2m turbine which is rated at 5kW.

As a consequence of the low energy extraction rate and the large amount of material, the embodied energy to energy production ratio is also relatively high, as is the payback period.

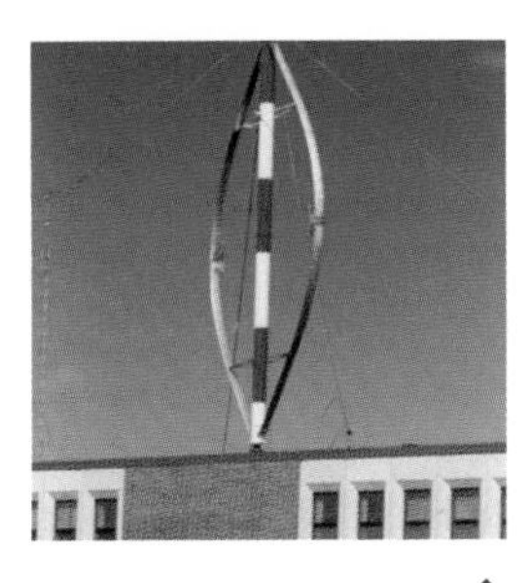

Figure 4.2 Twin-blade DAF-Indal Darrieus integrated into Canadian school circa 1980 (NRCan)

The original twin-bladed Darrieus type (commonly referred to as 'egg beater') has been trialled in several large-scale wind farms. It has even been integrated into a building as depicted in Figure 4.2. Unfortunately, for this building-integrated turbine, the twin-blade configuration produced peak energy pulses twice each revolution leading to a rapid onset of building structure damage and prompt decommissioning.

These Darrieus type turbines have their generators at a lower level which makes for easier maintenance. However, having the blades closer to the ground has energy yield implications. These turbines also require a burst of energy (e.g. from the grid) to start the blades turning when a designated anemometer detects winds above the cut-in wind speed.

Figure 4.3 H-Darrieus type VAWT (lift type) as found on the Technisches Rathaus in Munich, Germany

There are several small companies now manufacturing the three-bladed H-type Darrieus (commonly referred to as Gyromills) as integrated into the design of the Technisches Rathaus in Munich, Germany. These three-bladed turbines have much smoother torque profiles, lower cyclic stresses and in some cases can be self-starting.

A particularly 'elegant' turbine is the vertical axis turbine manufactured by Quiet Revolution, London, shown in Figure 4.4. Here the three blades are twisted into a triple helix design which further smoothes torque profiles whilst adding to the aesthetic appeal. This 'QR5' turbine is 5 x 3m and is rated at 6kW (although a 12 x 6m version has also been proposed). Light emitting diodes can be incorporated into the blades to produce varied effects presenting the option of creating a video screen. Although this turbine is called 'quiet revolution', noise emissions should still be carefully assessed.

Figure 4.4 Quiet Revolution lift force type vertical axis wind turbine installed at Maryport Visitors Centre, Cumbria (Quiet Revolution)

Some VAWT enthusiasts have experimented with designs that aim to combine separate components that generate drag and lift forces. However, this is not assisted by the fact that optimum operation speeds for both types are not complementary. As VAWTs are usually more expensive than HAWTs the payback times are often higher. This is elaborated on in Part 3.

2) GENERATOR TYPES

All turbines have a generator. This is either connected directly or via a gearbox to the main shaft and converts the applied torque into electrical energy by the relative movement of a length of coiled wire in a magnetic field. Several different generator types are available on the market. Although this is a complex subject, a brief overview can be useful to assist with equipment selection.

Synchronous and induction

Alternating current (AC) generators come in two varieties: 'synchronous' and 'induction' (asynchronous). These differ in the arrangement of the magnets on the rotor (the winding of wire on the fixed 'stator' which surround the rotating magnetic rotor are usually the same in both cases). Less expensive induction generator types are less efficient and lose energy through the generation of significant amounts of heat (e.g. ~3 per cent during full operation). Induction type generators can cause problems when large quantities of wind energy are feeding into the grid as they suffer from voltage instabilities. Synchronous type generators are used in conventional fossil fuel power generating plants. Direct current (DC) generators (commutators) are the other main type of generator.

Fixed-speed and variable-speed

Fixed-speed turbines use the less efficient induction generators. Variable-speed turbines can use the more efficient synchronous generation as they are de-coupled from the grid by the use of solid state frequency converters.

Gearbox and direct-drive

The drive connecting the main rotor shaft to the generators can either be routed through the more traditional gearbox system (to increase the rotor speed) or a direct-drive (gearless) system. Direct-drives do not have the mechanical noise associated with a gearbox and so are favoured when noise emissions are a key consideration.

Turbine manufacturers have a tendency to design gearboxes to match the design speed of 'off-the-shelf' induction generators. These gearboxes will tend to convert typical large-scale turbine blade rotation of 50rpm to 1000rpm for six-pole induction generators and 1500rpm for cheaper four-pole induction generators (at 50Hz).

Generators with their magnetic rotors rotating at lower speeds can still produce the same required 'grid compatible' electric output if the length or the diameter of the generator is increased. This is because the power output is not only proportional to the strength of the magnetic field and the rotor speed, but also to the length of the generators and the square of its diameter. As the diameter has more effect than the length it is the diameter that is usually increased. Direct-drive turbines are often seen to have wide nacelles (as shown in the Enercon and Lagerwey turbine images presented throughout this text).

As lower rotor speeds require larger generators they use more material and so are more expensive and heavier. However, the lower-speed designs increase simplicity and reliability, reduce the need for maintenance and extend the life of the machine. To illustrate this point it may be useful to consider that the rotor in the generator of a standard large-scale gearbox induction turbine will have rotated in a few months of operation the same amount as a large-scale direct-drive turbine will have over 20 years.

In the large-scale direct-drive market, Enercon (Germany) dominates. However, recent contenders have been Vensys (Germany), Jeumont Industrie (France) and MTorres (Spain). The Lagerwey (Netherlands) direct-drive technologies are now developed by Zephyros and Emergya Wind Technologies (EWT). In the small-scale wind energy markets most turbines are direct-drive.

Permanent magnets and electromagnets

For large-scale turbines it is common to have the magnetic field provided by electromagnets (which use electricity to generate a magnetic field). Most small-scale turbines use permanent magnets.

Despite their name, permanent magnets will lose the strength of their field if the iron dipoles are misaligned ('degaussing'). Therefore, permanent magnets should not be dropped (excessively agitated) or overheated (e.g. through the heat generated by induction generators). Power from permanent magnet turbines can be monitored over time to detect any decay in performance at a fixed wind speed.

3) BLADE DESIGN

Coefficient of performance

The aim of a blade manufacturer is to get as close as possible to the theoretical limit of extractable energy from the wind, which Albert Betz calculated in 1926. The Betz limit, or maximum possible coefficient of performance C_p, is 59 per cent (or 16/27) of the total available wind energy passing through the swept area of the blades.

The three-dimensional form of HAWT blades that have been optimized to a certain degree for maximum performance is complex. For large-scale HAWTs, the blades will twist to track the 'direction' of the 'apparent wind', which varies considerably along the blade as the distance from the central hub increases (see Box 4.2).

Once the complexity of blade design begins to be understood the amount of research a manufacturer is required to carry out can be appreciated. Some smaller companies are not able to invest in extensive research and as a result the full potential for the coefficient of performance of their blade may not be fully exploited.

Coefficient of performance for differing number of blades and tip speed ratios (TSR)

The common-sense idea that increasing the number of turbine blades will increase the power a turbine can extract is correct. However, when dealing with 'lift' turbines, the relationship between power and number of blades is not as strong as one may naturally assume. For example, a two-bladed turbine can produce almost the same energy as a three-bladed turbine if another factor is taken into account: the tip speed ratio. The tip speed ratio (TSR) is simply the ratio of the speed of the blade tips to the speed of the wind which is causing the blades to turn. The maximum TSR of simple drag-based devices is around 1 (a cup anemometer, for example, cannot have a cup moving faster than the speed of the wind it is catching). The TSR of lift-based turbines is usually around 6 or 7, which results in what is commonly termed high 'solidity'.

A two-bladed turbine with an appropriately elevated tip speed ratio will have a similar 'solidity' to a three-bladed turbine with a lower tip speed ratio. One-bladed turbines have been designed

BOX 4.2
THE VARIATION OF APPARENT WIND ON LARGE-SCALE HAWT BLADES

The 'apparent wind' is the resulting wind vector a blade experiences based on the wind speed and the local blade rotation speed. For a HAWT the local blade rotation speed will increase in the direction of the tip of the blade. Therefore, the direction of the apparent wind will vary across the blade. This is the reason large well-designed blades possess a degree of curvature in order to maintain the optimum pitch angle in line with an optimum constant 'angle of attack'.

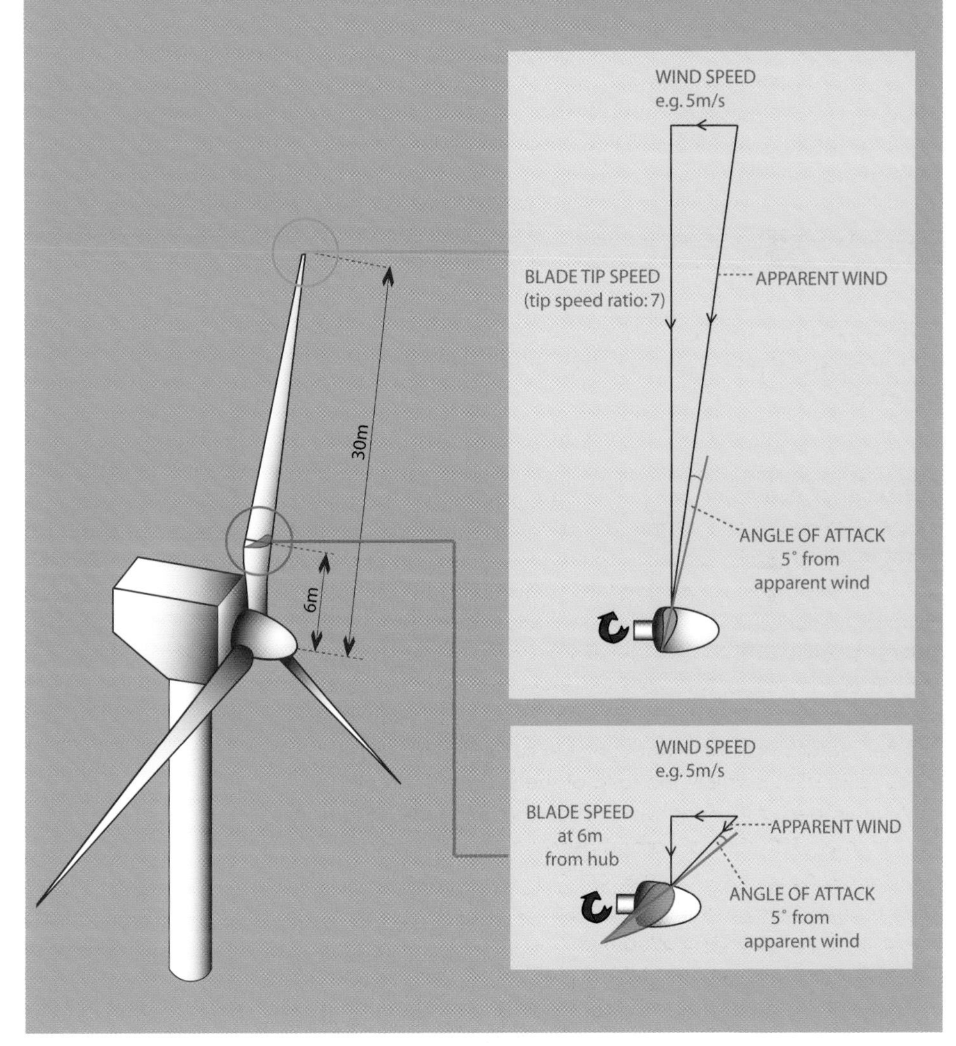

and successfully operated. Although they use less material and may be considered to have a quirky aesthetic, they now exist only as historical curiosities.

The graph in Figure 4.5 shows the relationship between tip speed ratio, the number of blades, the lift to drag ratio, and the overall coefficient of performance for a notional blade design. Although turbines running at high TSR will have higher Cps they will also create higher levels of audible tip-generated turbulence. As higher speeds tend to increase drag (which decreases the performance), many blades tend to have a very thin profile in order to reduce the drag component and increase the lift to drag ratio. Box 4.3 illustrates how the tip speed ratio can be calculated.

Several two-bladed turbines are on the market, e.g. from Vergnet and the Scirocco from Eoltec (Figure 4.6). The Scirocco has a blade diameter of 5.6m and is rated at 6kW at 11.5m/s (which remains constant up to 60m/s). Eoltec have opted to keep the TSR close to 6 (245rpm) in order to keep tip speed relatively low (peak around 70m/s) to reduce sound emissions and blade wear.

Large-scale HAWTs have a lower rpm and may appear to rotate relatively slowly. However, if the TSR is 7, then during 12m/s winds the tips will be travelling at speeds approaching 90m/s. Even during mean wind speeds of 6m/s the tip speed will be over 40m/s. Considering these facts may help generate an appreciation the origin of some of the aerodynamic noise emissions, as well as the potential result of an encounter with the blade, e.g. a bird strike.

Figure 4.5 Variation of coefficient of performance (C_p) with tip speed ratio (TSR) for different numbers of blades for HAWT for blade designs with a lower quality blade design (lift to drag ratio, LDR = 60) and a higher quality blade design (LDR = 160)[1]

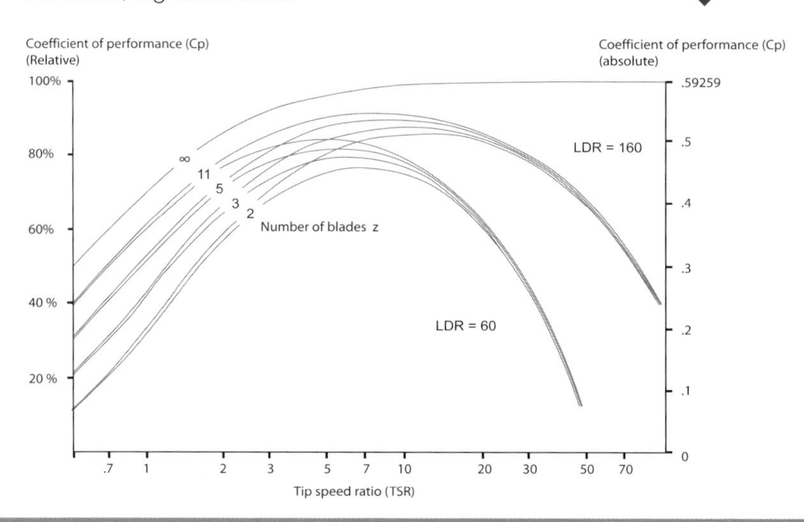

Figure 4.6 The *Scirocco* from French manufacturers Eoltec, 5.6m turbine rated 6kW at 11.5m/s (www.eoltec.com)

Coefficient of performance and wind speed

As the coefficient of performance, Cp, will vary widely between manufacturers, performance data should be obtained wherever possible direct from the manufacturer. Throughout this text, 30 per cent is taken as a general average approximation for Cp. However, in reality Cp will vary with wind speed (as shown in Figure 4.7).

One company deserving mention for efforts in blade design is Enercon. In 2004 Deutsche Windguard Consulting recorded their latest turbine having an overall efficiency of 0.53 which translated into an aerodynamic efficiency of 56 per cent when applying the 95 per cent generator efficiency.[2] This figure is very close to the theoretical 59 per cent limit. Their literature shows their turbines achieving an overall Cp of ~0.5 for the most important wind speeds of 7–9m/s across their range of turbines.

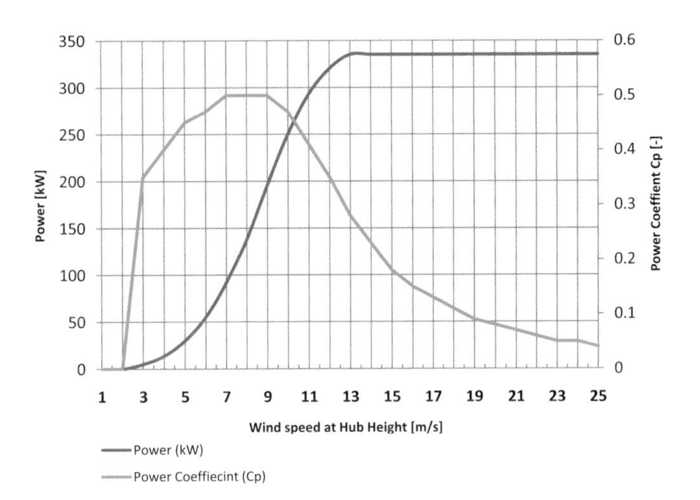

Figure 4.7 Performance data for the E-33 HAWT with 33m blade diameter, rated at 330kW with a TSR of 6, from the German manufacturer Enercon (www.enercon.de)

BOX 4.3
CALCULATING THE TIP SPEED AND THE TIP SPEED RATIO

The TSR is rarely given by manufacturers. However, it can be readily calculated using the maximum rotational frequency (rpm), which is often quoted.

1) Convert the maximum rotation speed f from rpm to rps (60rpm = 1rps)
2) Calculate the angular rotation speed ω ($\omega = 2\pi f$)
3) Calculate tip speed v_T using the blade radius r ($v_T = r\omega$)
4) Calculate the tip speed ratio TSR using the rated wind speed v_R ($TSR = v_T/v_R$)

4) PROTECTION

The energy content of high speed winds will cause excessive wear on turbine components if preventative measures (e.g. to slow or stop the blades turning) are not taken. Without 'overspeed protection' severe winds would cause a catastrophic failure. For this reason all lift-based turbines will have mechanisms to prevent damage to a turbine at high wind speeds.

Turbines should also be designed to stop automatically if failures occur, e.g. if generators overheat or if the turbine is disconnected from the grid (which causes a rapid acceleration of the blades as they are allowed to run free without resistance provided by active generation).

Two separate fail-safe systems are generally selected from a wide range of possibilities. The primary system can involve one of the following:

Passive stall
This technique involves designing the blades to cause the air to separate from the foil naturally when the wind speed reaches a certain speed (see Box 4.4). Although a reliable safety mechanism, the blade design will result in energy losses, especially at high speed (above the rated speeds) where the generated power output will tail off very quickly as the lift forces diminish.

Active blade pitch control (also called 'feathering' or 'furling').
This involves a mechanical means of rotating the full length of blade (or a proportion of the blade) radially. This is a very reliable means of breaking by eliminating the lift forces and it does not cause major equipment stresses. It is the standard means for overspeed control in large-scale machines, and modern turbines will be able to pitch each blade individually in order to optimize blade efficiency.

BOX 4.4

PASSIVE STALL BLADE DESIGNS FOR OVERSPEED CONTROL

With passive stall designs the blade is fixed and the 'angle of attack' is specifically set to be relatively high. During low wind speeds the air is able to flow over the blade and remain 'attached' to the surface and thus generate 'lift' forces. However, at high wind speeds (e.g. 12m/s) the air will begin to detach or 'separate' from the blade and consequently the lift force decrease rapidly as depicted below.

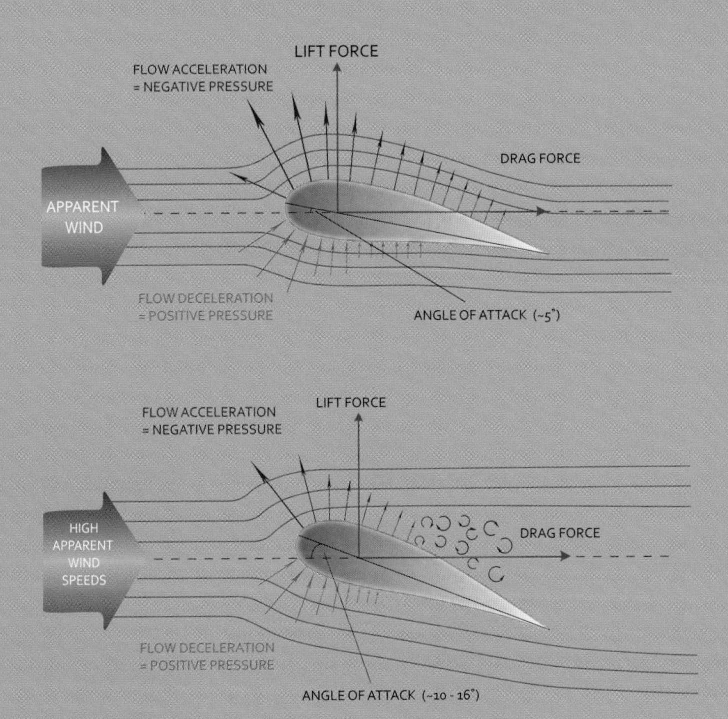

During wind speeds above the rated value very little energy will be generated by passive stall turbines. Although these higher wind speeds are less frequent they do have a high energy content. Active pitch blade control allows the blade to rotate and adjust to keep the energy produced at these high wind speeds constant – i.e. at the rated peak kW.

Yawing and tilting

HAWTs are designed to face (or yaw) into the wind – either actively in the case of large-scale turbines or passively in the case of most small-scale turbines (via a tail fin). Turning the blades out of the wind will slow the blades. This technique is not used with large-scale machines as yawing out of the wind during high speeds causes undue component stresses which can lead to equipment failure.

Figure 4.8 Tip breaks that can be activated during high wind speeds on an Entegrity Wind EW50 (15m/50kW)

Blade bending

Several small turbines allow their blades to bend with the wind to decrease the lift force in high wind speeds (e.g. Proven turbines).

Tip control/breaks

In some turbines tip breaks (fins) can be activated to reduce the speed of the turbines (see Figure 4.8). These can be spring/hydraulically operated to allow them to work in the event of an electrical power failure.

The secondary system will be one that is less advisable to use on a regular basis to control blade speed, e.g. a mechanical brake (which would wear given regular use) or applying a current to the generator (which can cause the generator to burn out). In addition, a mechanical locking system is required to lock the turbine during maintenance.

Figure 4.9 Corporate identity – wind turbines at a supermarket in Greenwich, London (1999) (Russell Curtis)

Lightning protection

The electrostatic discharge between clouds and the Earth has potential voltages up to 100MV and peak current between 2kA and 200kA. A lightning strike to a turbine will generate large electromagnetic mechanical forces between adjacent conductive materials. Although the risk is small, lightning has hit turbines in the past causing major damage.

The chances of a strike can be estimated. For example, British Standard BS 6651 provides a means to predict a strike risk factor for a building or structure depending on geographical region, terrain, the size and material of the structure. Therefore, when appropriate, turbines should have robust lightning protection systems (LPS).

5) TOWER DESIGN

Figure 4.10 Turbine viewing gallery at the Ecotech Centre, Swaffham, UK

Smaller-scale urban turbines are most commonly mounted on steel towers. Large-scale turbine towers are also usually made of steel although concrete towers (or steel/concrete hybrid) are being considered more as tower heights increase. Lattice towers or cable (guy) supported steel towers can be used. However, in the case of the lattice towers, although they can be more cost efficient, the aesthetics may not be considered appropriate. In the cases of guy supported towers space may be an issue (as guys will extend for considerable lengths in several directions).

In public areas, lattices can be climbed and guy cables can cause obstruction or be tampered with by individuals. As discussed in the environmental impact section, tower designs should not only be secure but look secure. This usually means thicker towers and there are cost implications for this aspect of urban tower design.

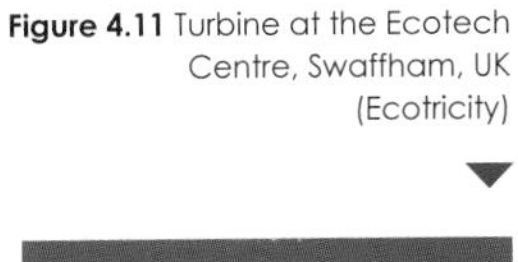

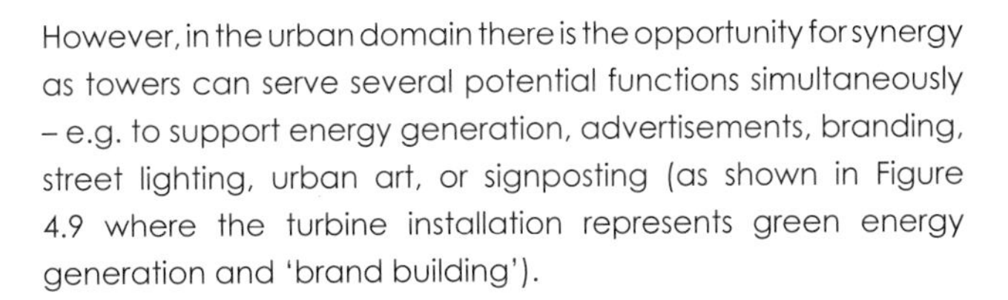

However, in the urban domain there is the opportunity for synergy as towers can serve several potential functions simultaneously – e.g. to support energy generation, advertisements, branding, street lighting, urban art, or signposting (as shown in Figure 4.9 where the turbine installation represents green energy generation and 'brand building').

Figure 4.11 Turbine at the Ecotech Centre, Swaffham, UK (Ecotricity)

There are a small number of turbines with public viewing galleries integrated into the top of the tower. For example at the Ecotech Centre in Norfolk, UK, the viewing gallery allows the public to be fully immersed with the technology if they are up to climbing the 67m tower via 300 steps (See Figures 4.10 and 4.11). The turbine was installed In 1999 (an Enercon 66m) and this 1.5MW turbine is reported to generate enough electricity to power around 1000 homes. The turbine is 360m from the nearest residential houses and there have been no noise-related complaints since its operation.

How high should a tower be?

Whether in open areas or in urban areas, higher towers are almost always of benefit in order to allow a turbine to reach higher quality, energy-rich winds. The height of the tower is even more of an important concern in the urban area. The question of tower height therefore goes hand in hand with wind resource assessment, as discussed in Part 3. As an absolute minimum, a tower should be high enough to elevate the turbine out of turbulent zones which prevent a turbine extracting the wind energy (see Box 4.5).

In some cases the question of tower height is answered by either planning constraints or the manufacturers, who will often only offer specific 'standard tower' heights, although some will offer a range.

Figure 4.12 6kW Proven turbine being erected on the roof of Ashenden House, London (2007) (Photon Energy)

BOX 4.5

THE NEGATIVE EFFECTS OF TURBULENCE ON ENERGY YIELDS

The swirling, turbulent nature of low level winds, found in many urban areas, has two main consequences that reduce the amount of energy captured by a lift-based turbine (e.g. horizontal axis wind turbine):

- The turbine will constantly be attempting to turn (yaw) into the wind and so reduce the amount of time it is aligned with the flow.
- Even when aligned with the main wind flow direction, the air will arrive at the blades at non-optimal direction for the blade design.

The top image shows the idealized case where the wind has a low turbulence intensity – as can be found in wind tunnels (e.g. during turbine testing). The bottom image depicts the transient swirling nature of wind with a high turbulence intensity. During these types of flow the deviation from the idealized flow can cause flow separation which reduces the lift force and the energy the turbine can capture.

Drag type vertical axis wind turbines are much more resilient to these phenomena. Lift-based vertical axis wind turbines will be less susceptible to losses through the first mechanism; however, they will still be subjected to decreased performance due to the second principle.

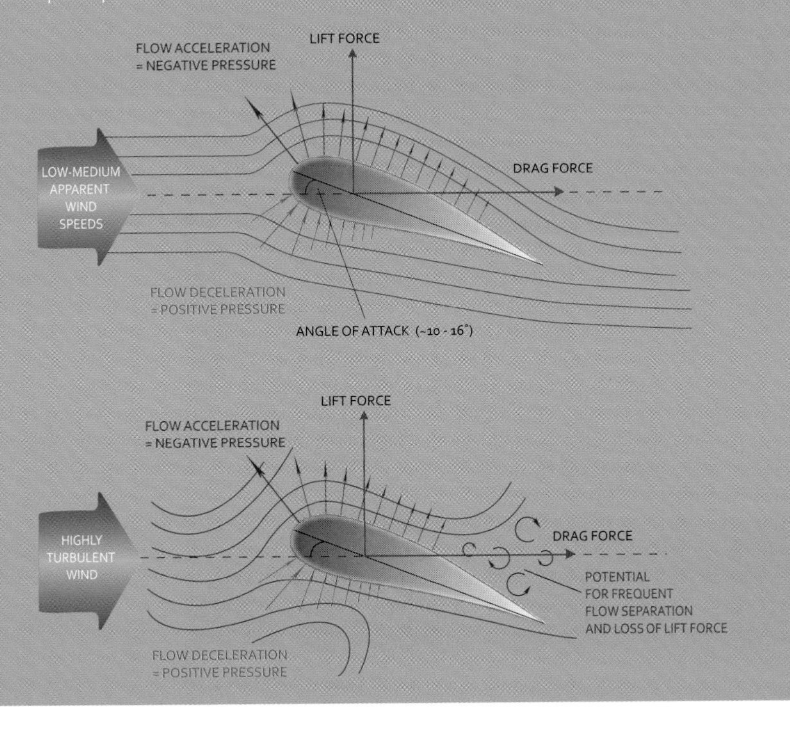

Erecting towers on building roofs

Sizeable towers can and have been erected on suitable roofs of buildings. One example is the 6kW Proven turbine installed on an 11-storey residential building, Ashenden House, in Elephant and Castle, London.

Practical considerations such as a suitable supporting structure and access for a crane are important. Space for winching may also be needed, as shown for the 9m tower in Figure 4.12. The installer, Photon Energy, have indicated there have been no reported vibration or noise issues.

Figure 4.13 Ontario Electrical Construction Company visibly supporting wind energy (Harold L. Potts)

6) GRID CONNECTION

As battery storage tends to be expensive, and typical lead-acid batteries have to be replaced after five to eight years (creating future cost implications), grid connection tends to be the preferred option for most urban turbine installations. To connect to the grid the correct equipment must be installed. Generators typically generate at 3 phase which is then rectified to DC before being converted to mains-level voltage by an approved inverter. Connecting to the grid will require different procedures depending on the country. For example, in the UK turbine installations require:

- Grid connection permission. Permission to export to the grid can be made available by the local distribution network operator (DNO). A charge may be made for this service depending on a number of issues such as size and voltage of the turbine. This will be subject to a subsequent inspection of equipment such as protection relays and the export meter (which must be installed to measure the amount of electric that is being exported).

- Sales agreement with an electricity supplier. Since the turbine will be generating electricity at night and the development may not use the entire load (depending on the size of the turbine and any planning conditions relating to noise), it is preferable to be able to sell electricity to a supplier (this does not have to be with the local DNO). The deals available vary depending on how much electricity can be exported and on the supplier.

There are several schemes, such as 'net metering' and 'feed-in tariffs', implemented in certain regions to encourage local renewable energy generation.

Standard grid connections allow the owner of renewable energy technology to reduce their electricity bills by using their own electricity on-site instead of buying in electricity from the grid. Providing enough electricity is being used at the same time local electricity is being generated, this scheme is equivalent to being paid at the 'kWh buy-in rate'. Any surplus electricity will go into the grid. However, the rate per kWh of this exported electricity is usually much lower than the buy-in cost, although some electricity companies have voluntarily matched the buy-in charge.

'Net metering' schemes provide additional incentives as they allow the owner of renewable energy technology to reduce their electricity bills by consuming their own electricity on-site and, importantly, use any surplus electricity exported to 'turn back' their electricity meter. If more energy is exported than consumed in a certain location, over a given time period, a credit may be negotiated.

A 'buy-back/feed-in tariff' requires all the power generated on-site to be fed directly into the grid. Under this scheme, the utilities are obligated, by the government, to pay a fixed price (or 'tariff') per kWh generated. Like net metering schemes, feed-in tariffs are used to stimulate the market for certain technologies. However, they can be a more effective stimulus as they can pay more than buy-in rate. However, this scheme goes against the idea of encouraging electricity producers to use their own electricity on-site, which is more efficient as the transmission losses are less.

In the UK, Renewable Obligation Certificates (ROCs) are used as a means to provide incentives and allow the use of locally generated energy on-site (see Box 3.4, page 105).

Grid connection companies may be forgiven for seeing turbines as unwanted complications, but some are at least seen to be embracing renewable technology. The trend should continue as more and more generation companies wake up to the change.

The specifics of grid connection will depend on the turbine being considered and engineering schematics should be readily available from all manufacturers.

It should be noted that, counter to intuition, standard small wind (and PV) energy systems (i.e. those which are not part of organized decentralized distribution networks) will not provide energy when there is a power cut in the grid. This follows from standard safety regulations to prevent possible injuries to those who may be working on the grid system following a power interruption.

7) SOURCING EQUIPMENT

There is a considerable distinction between the small-wind market and the large-wind market. In the larger turbine market, the similarity of many (although not all) design features makes sourcing of large turbines a relatively straight-forward process. However, in the past the 'lead-in' times for delivery of a large-scale turbine have increased with increasing demand for wind energy and manufacturing bottlenecks related to component supply such as generators and bearings.

Sourcing of small turbines can be a more involved process. Determining the most appropriate technology for a given project can be difficult due the wider range of options. Adding to the complexity is the wide variation of turbine quality/performance. The advancement in design seen with large-scale turbines has not entirely filtered down to the smaller turbines (e.g. having individual automated pitch control on blades) due to physical and financial constraints. However, the most appropriate turbines for a given project can be identified with a moderate amount of investigation, and a list of manufacturers is given in Appendix 1.

The most difficult category to source is the mid-range turbine market. The mid-range market is particularly relevant to urban wind turbines as there may be planning constraints (such as visual impact and noise) or other constraints related to budgets in wind energy co-operatives. However, the number of options available in this niche market is dwindling as the gap between large- and small-scale turbines is widening. The recent development in technology and a drive to create machines that generate more power have seen the majority of the main wind turbine manufacturers moving away from making new turbines with blades that have a diameter of 20 to 50m (for example with a rating between 100 and 600kW) in favour of concentrating on larger megawatt turbines.

The main large-scale manufacturers include: Vestas (Denmark), GE (US), Enercon (Germany), Gamesa Eolica (Spain), Neg Micon (now part of Vestas), REpower (Germany), Nordex (Germany), Suzlon (India), Acciona (Spain), Ecotecnia (Spain), Siemens (Germany), MHI (Japan), DeWind (Germany), Clipper (US) and Goldwind (China). The smallest turbines available from some of the largest manufacturers are given in Table 4.1. As can be seen, most are now megawatt turbines.

Turbine manufacturer	Turbine model	Blade diameter (m)	Rating (kW)
Suzlon	S.30.350 kW	30	350
Enercon	E-33	33	330
Goldwind	43/600	43	600
Ecotecnia	48	48	750
Gamesa	G52	52z	850
Vestas	V52	52	850
MHI	MHI-1000A	57/61.4	1000
Nordex	N60	60	1300
Siemens (Bonus)	SWT-1.3-62	62	1300
REpower	MD77	77	1500
GE	GE 1.5s	70	1500
Acciona (EHN)	AW 70/1500	70/77	1500

Table 4.1 Smallest turbines currently available from the main large-scale manufacturers (2008)

As the industry develops there is will be a tendency for these smaller models to be superseded by larger models unless a manufacturer deliberately chooses to address the urban/community wind market. However, some manufacturers are beginning to embracing this mid-range market. There are other smaller manufactures producing mid-range turbines which can also be investigated (see Appendix 1).

Wind class

Commercial wind turbines are designed/optimized with certain wind speed and turbulence conditions in mind and these have been categorized into recognized standards by IEC (wind turbine standards IEC 61400) as shown in Table 4.2.

As many inhabited areas have lower wind speeds the appropriate lower wind classes should ideally be selected. These turbines will have a lower turbine rating than their higher wind class counterparts (with the same blade diameter). However, they will tend to produce more energy at low wind speeds than a higher wind class turbine.

Gamesa produce a lower wind class turbine (G58) rated at 850kW and Suzlon have a 52m diameter blade turbine rated at 600kW (S.52.600) designed especially for low to medium wind speeds.

Table 4.2 International Engineering Consortium (IEC) wind turbine classes

	Class I	Class II	Class III	Class IV
Mean wind speed	10m/s	8.5m/s	7.5m/s	6m/s

Re-engineered turbines

The move by manufacturers from mid-range to megawatt turbines has been accompanied by a tendency for wind farms built several years ago to be re-powered. Re-powering involves replacing smaller turbines with larger, higher capacity turbines. As the life expectancy of wind turbines is usually at least 20 years, the turbines that have been replaced are often fully functional and capable of running for many years to come. This move to re-powering has led to the development of a new market for re-engineered or 'second-hand' turbines (as shown in Figure 4.14).

Figure 4.14 A Re-engineered Micon M350 (250kW 26m blade diameter)

A number of second-hand turbine dealers have emerged from which these mid-range turbines are available. Vestas are a common second-hand mid-range make with a variety of models available with different blade diameter – e.g. V47 660kW (blade diameter 47m) as shown in Figure 4.15.

As these turbines are older it will be particularly important to assess noise generation and the 'availability' (reliability) thoroughly. They can be a good investment for communities who may wish to see their co-op as a multistage investment with the first stage being a re-engineered turbine and additional stages replacing the turbine, after several years with a new and larger turbine after energy generation and low environmental impacts have been proven.

Figure 4.15 Vestas V47, 47m blade diameter, which may be available as a re-engineered turbine

There may also be the option to fully service this re-engineered turbine and reposition in a new location in the same neighbourhood.

Sourcing for building-integrated turbines

Sourcing for building-integrated turbines can be difficult. Manufacturers can be forgiven for being less than enthusiastic in their response to requests for information related to this topic. Even if they are keen in principle, there are several good reasons for a manufacturer, not to undertake such a project:

- They may be understaffed and overloaded with 'normal' and less involved enquiries.
- They may have had several enquiries from uninformed individuals on this subject which have 'gone cold'.
- The capital return for the associated effort may be lower than for more standard projects.
- The capital return for the perceived risk will be lower than for more standard projects.

Fortunately there are several good reasons for manufacturers to engage with building integrated wind energy projects:

- They will be creating a very visible statement – which, if designed correctly, will generate confidence, interest and investment in their market and company.
- They will be considered leaders in an emerging market.
- They get the opportunity to work on an interesting and challenging project.
- The increase in effort will be spread out over a fairly long period and so not impact excessively on day-to-day activities.
- The risks can be minimized by ensuring the team agree to follow fundamental engineering principles.
- They will be advancing a path into a sustainable renewable technologies future.

Aside from direct revenue, the indirect benefits from engaging in these projects can be significant. The value of a company can be increased by expanding their project portfolio, improving staff retention, expanding the market and increasing market visibility. For example, Danish manufacturers Norwin have elevated their profile with their involvement with building-integrated turbine projects such as the Dubai World Trade Centre, Castle House (UK) and The LightHouse (Dubai).

In the first instance, manufacturers may respond more favourably to 'non-standard' requests if the enquirer has some demonstrable grasp of the subject and has the visible backing from an experienced/committed team.

8) WIND ENERGY YIELD ENHANCEMENT TECHNIQUES

When the subject of wind enhancement is approached, visions of aerodynamic shapes accelerating prevailing wind into turbines may come into mind. However, there are several fundamental areas that should also be covered before getting into the methods of making use of physical surrounds and obstacles to enhance winds. The following points summarize subjects discussed throughout this text regarding overall wind energy yield enhancement:

Ensure adequate resources:

- careful site and tower selection – e.g. low turbulence winds;
- monitor on site, ideally for one year;
- ensure the annual mean wind speeds exceed 5m/s at hub level.

Maximize swept area:

- energy doubles as swept area doubles.

Supply local buildings:

- Lower transmission losses and costs;
- sell to local buildings at a 'healthy' rate.

Use appropriate turbine design:

- ensure good Cp at wind speed relevant to site (analyse Cp variation with wind speed);
- ensure good generator efficiency.

Reliability:

- use suitable technology – e.g. low speed generator, good overspeed control;
- seek manufacturer guarantees;
- ensure budget for maintenance;
- complete comprehensive environmental impacts assessment.

Wind acceleration and turbine siting

Onshore wind farm developers purposely targeted sites where natural acceleration occurs, such as hill tops or slopes up from coastal regions, and where annual mean wind speeds are high. Incidentally, these regions often coincide with areas of outstanding natural beauty and inappropriate choices can often be made where environmental impacts are not given sufficient consideration. For urban wind energy, the use of these wind-accelerating topographical features is encouraged and these large-scale features should be kept in mind to help ensure energy production is high.

Generally, in urban locations, obstructions such as buildings tend to lower wind speeds and increase atmospheric turbulence. Any resulting local wind enhancement is often unplanned and unwelcome. Pedestrian discomfort often occurs in areas near tall buildings where the wind can impinge on a building and flow down the façade (downdraught) or accelerate between buildings (funnelling).

Artificial wind enhancement techniques – via the use of movable or non-movable structures (e.g. buildings) – seek to optimize the same physical phenomena which give rise to the hill and funnel effects in nature. It should be noted, however, that unless a building is specifically designed with wind turbines in mind it is unlikely that there will be any significant positive acceleration benefits. The likely result will be a noticeable drop in energy yields.

Nevertheless, it is possible to plan a development to enhance wind speeds in particular areas by combining buildings and landscaping to create artificial tunnels, hills and embankments for placing wind turbines. Therefore the subject of wind enhancement is readdressed when we consider wind enhancement through purposeful building design, which is discussed (and to some extent quantified) in Part 5 when building-integrated turbines are examined.

It should be noted that pedestrian comfort/safety should also be kept in mind when attempting to deliberately accelerate winds, as low wind speeds are often desirable along walkways and in public spaces.

SUMMARY

A good grasp of turbine fundamentals is necessary in order to select the appropriate technology for any given project.

For large-scale megawatt turbines only lift-force HAWTs are available although there are specifics relating to blade and generator types which are important to bear in mind. The same applies to medium-scale turbines although the choice is limited to a few well-established manufacturers for new machines or to re-engineered turbines which come onto the market as wind farms are upgraded.

In the small turbine market, the situation is more complex with the availability of a wide variety of lift-based HAWTs and both lift and drag VAWTs. In addition, the variation of reliability and performance of the turbines on the market is expected to be substantial.

Sourcing of equipment for building-integrated turbines can be time consuming. However, this should not be overly limiting, and sourcing should begin as soon as possible in order to provide clarity on the available options for a given project.

The variation in blade performance depends on several key factors such as tip speed ratio and the lift to drag ratio. A well-designed blade/generator can reach efficiencies around 50 per cent, which is close to the theoretical maximum 'Betz' limit of 59 per cent.

Other design elements which should be considered include the 'overspeed control' method which protects turbines from high wind speeds. There are several differing methods and these mechanism affect not only the safety and longevity of a turbine but often the energy performance and cost.

The ease and length of the grid connection process will depend on the region in which a turbine is being installed. The trend of easier grid connection should continue as more and more power companies wake up to changes in energy production and begin to embrace decentralized and local renewable energy generation.

Techniques to ensure enhanced wind energy yields begin with assessing natural topographical features such as hills, slopes or coastal fronts and ensuring the fundamental areas of wind energy generation are covered. These include: ensuring comprehensive monitoring of wind resources, maximizing swept areas, supplying local buildings, using high performance blades and generators and ensuring good turbine 'availability', which includes assessing environmental impact to ensure the turbine is allowed to run once erected.

Wind enhancement through purposeful building design, and other issues relating to building-integrated turbines, are discussed in the next and final part.

Figure 4.16 Green Park Turbine, UK (Ioannis Rizos)

REFERENCES

1 www.ifb.uni-stuttgart.de/~doerner/edesignphil.html (Dec 2008)

2 Deutsche Windguard Consulting Gmbh, www.windguard.de (Jan 2009)

Building-Integrated Wind Turbines

The first large-scale building-integrated turbine project – World Trade Centre in Bahrain, 2008 (Ahmed Hussain)

INTRODUCTION:
POWERFUL ARCHITECTURE

Building-integrated wind turbines are associated with buildings designed and shaped specifically with wind energy in mind. The larger the scale of the turbine(s) proposed, the greater the environmental impacts on both the surrounding environment and the building itself, and the greater the challenge of designing and constructing a building that also meets the needs of its owner(s) and occupants becomes.

It is not just a matter of the number, scale, type and location of turbine(s), predicted annual energy yield and design life. Integration of these dynamic rotating machines can influence decisions on building orientation, massing (form and height), local façade design and curvature, structure (loads and vibration), acoustic isolation, choice of natural/mechanical ventilation, spatial layouts, access for maintenance, safety features, electrical services design, construction techniques, commissioning, replacement/decommissioning, whole life costs, and so on.

The design process can absorb significant time and (specialist) resources, so care needs to be taken that other aspects of the design that impact greatly on operational energy demand (e.g. level of insulation, glazing, solar shading, ventilation, daylight penetration, material use, waste, water cycles etc.) are not neglected, enabling a holistic sustainable building design to be realized, i.e. one that is buildable and actually delivers excellent performance in operation over the design life of the building.

Small wind turbines mounted on existing buildings will fall into the category of building-mounted turbines and retrofitting, covered in previous sections. However, those interested in retrofitting may still find some of the information contained here useful.

This section builds on Part 2, which first addressed building-integrated turbines. It will serve to inform decision-makers and designers on specialist topics over and above those associated with more standard wind energy projects. This includes evaluating the energy viability of a given building-integrated wind turbine using the latest simulation technology, as well as the more pertinent technological and practical issues.

This final section covers the following:

1. General guidelines and options for building-integrated wind turbines.
2. Wind directionality and building orientation.
3. Predicting energy yields from turbines integrated into shrouds within tall buildings.
4. Environmental impacts, building design and planning.

1) GENERAL GUIDELINES FOR BUILDING-INTEGRATED WIND TURBINES

Integration of one or more wind turbines into or onto a building or structure must overcome some fundamental issues to help ensure the success of the development:

- It is necessary to ensure that adequate wind resources are available. This may mean deliberately elevating the turbine to access better quality winds and/or accelerating winds using the building form (as the wind speeds in urban areas are generally lower than in adjacent rural locations due to the resistance caused by the presence of buildings and infrastructure).

- The static nature of a building should be considered in relation to varying wind directions. A stand-alone turbine is free to yaw (turn) into the direction of the prevailing wind in order to optimize power extraction and this should be taken into account, e.g. through building orientation and form. Therefore, an assessment of wind direction is required (as well as wind speeds).

- The shape and orientation of the building may be limited by the constraints of the site in many urban areas and this may reduce wind energy feasibility.

- Turbines used in wind farms are normally located a substantial distance (>500m) away from surrounding properties to ameliorate their noise, visual and safety impacts. Therefore environmental impact assessment should be given a priority (as discussed in Part 3 and later in the section).

- In some cases there may be a need for acoustic isolation on areas close to rotors. This should not bring about the requirement of forced ventilation (having regions with sealed windows) where natural ventilation could ordinarily be used (as there will be increased energy usage).

- Pedestrian comfort (e.g. in the public realm at the base of the building) should not be compromised when aiming to deliberately build in windy regions or accelerate winds. Several amelioration techniques can be used at ground level – such as canopies, screens and greenery (although the inherent form of the building is the most influential factor).

- Accelerating winds produce negative pressure on adjacent façades. These will work against ventilation intakes, although they will complement ventilation and smoke extracts and significantly reduce the need for fan power. Therefore intakes/extracts should be positioned accordingly. Accelerated wind may also affect window opening or external shading strategies and simple measures can be adopted to prevent issues (e.g. inward opening windows).

- Access for erection and tower maintenance should also be given consideration – i.e. whether access for a crane is required and where it can be positioned.

- The aesthetics of wind turbines should be given consideration to ensure it complements rather than contrasts with those of the building.

If substantial benefit is to be derived from the integration of wind energy into the built environment, the building(s), whose electricity demands the wind energy generation will partially meet, must be inherently energy efficient for any integrated wind energy generation to have a significant impact on the net energy balance of the building(s). Wind turbines integrated into the built environment should, generally, be capable of producing a significant proportion (say 10 per cent) of the annual electricity demand of the surrounding building(s). Otherwise, the turbine(s) will become a primarily aesthetic feature which would be unsatisfactory from both an environmental viewpoint and that of clients, planners, designers, occupants and local inhabitants. The potential for wind energy generation for the development proposed in Project WEB in terms of the total electricity demand is given in Figure 5.1.

Figure 5.1 Percentage of electrical energy supplied by the three 250kW turbines of the Project WEB twin tower as a function of building type and climate/terrain conditions in Dublin, Ireland (Project WEB Handbook – BDSP Partnership, 2001)

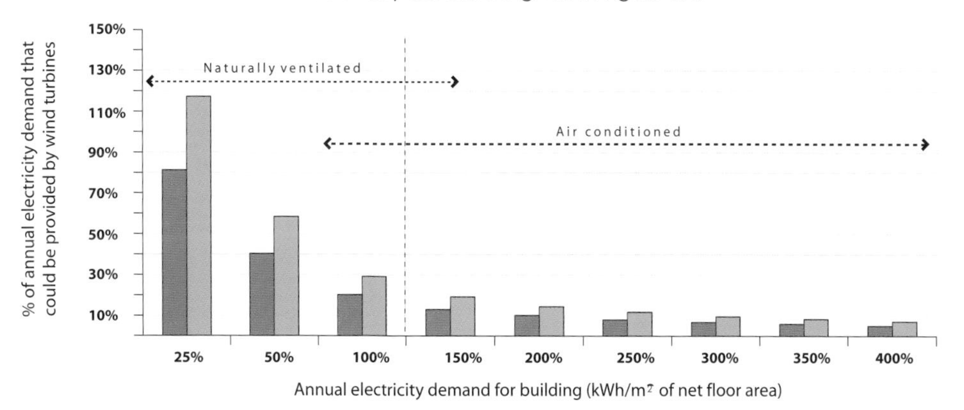

Although one of the most important factors of wind energy viability relates to the energy generation potential, when dealing with urban wind energy each case should be evaluated on its individual specific merits. This is especially relevant for building-integrated turbines as there could be a wide variety of potential development proposals.

The CH2 '6 Star energy rating' (Council House No. 2) office building built in Melbourne, Australia in 2006, is an example where wind energy was conceived as part of a holistic design system. The 6 drag type VAWT mounted on the roof (shown in Figure 5.2) are designed to aid air extraction. The efficacy of these bespoke devices will be lower than their lift-based counterparts. However, the success of wind energy in this example relies less on it meeting a substantial part of the total energy demand and more on the synergistic aspect of all the green design features.

Figure 5.2 CH2 building in Melbourne, Australia by DesignInc Architects

The entire envelope is conceived as an expression of biomimicry and includes high thermal mass, phase change heat store, automatic window shutters, solar hot water collectors, bicycle parking, photovoltaic cells, chilled water cooling system and chilled beams, evaporative cooling, co-generation plant, open and flexible layout, as well as the inclusion in the budget of education/demonstration, art, natural recycled materials and planted greenery. The additional costs of the sustainable features are reported to pay back within 10 years[1] and this does not factor into account increased workplace efficacy as a result of an inspiring environment. Therefore these particular building-integrated turbines are an example of where the role of visibility and communication are of key importance.

Of course it is possible and preferable for building-integrated wind turbines to have a substantial energy generation capacity as well as acting as a cultural statement, and a number of viable options for basic design arrangements exist (with some performing better than others).

Generic options for building-integrated wind turbines

Building-integrated turbines can have numerous forms. Some of the main types (Types A–G) are given below together with some of the practical considerations. Although these different options are shown for a similar building form there is considerable scope for variety and adaptation in line with development briefs.

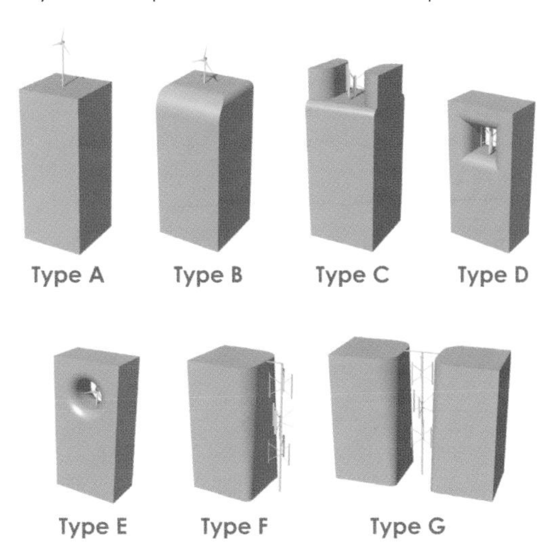

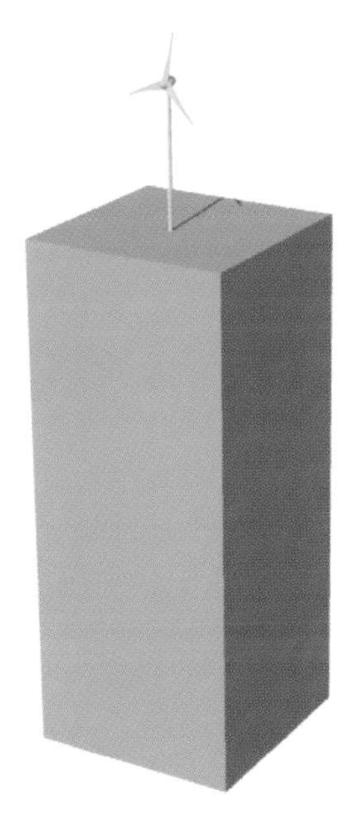

A) On top of building

This option primarily takes advantage of the opportunity to access higher-quality winds that tend to exist at greater altitudes. These winds will not only have a relatively high energy content but are also likely to be less turbulent. However, there will be a degree of natural wind acceleration (with a ~10 per cent energy increase for the natural wind acceleration alone – see Table 5.2). In general:

- The height of the tower is important to avoid the local turbulence envelope generated by adjacent sharp edges and also to ensure the blades are meeting air without any vertical component.
- A high tower will mean 'fall-over distance', tower erection, access for maintenance and vibration will be important considerations.
- High towers may mean planners will be concerned with visual impacts although it could be stated that the intention is to have the sustainability of the design made visible.

B) On top of rounded building

This option takes advantage of the higher quality winds at higher altitudes and additional local acceleration (with a ~15 per cent energy increase for the local wind acceleration – see Table 5.2). The rounded façade will mean the tower height can be much lower. In general:

- The extent of the rounding will influence the local acceleration. The additional costs for the façade may be offset by the increased value of the building from an improved character (aesthetics and green credentials).
- Low towers will mean lower visibility issues for planners.
- The lower tower will mean 'fall-over distance', tower erection, access for maintenance and vibration will be easier to reconcile.
- A lower tower will mean 'line of sight' for blade flicker and noise emissions will be improved.

C) Concentrator on top of rounded building

This option takes advantage of the higher quality winds at higher altitudes and notable local acceleration especially if the wind character of the region is bi-directional (~20 per cent energy increase due to local acceleration – see Table 5.2). In general:

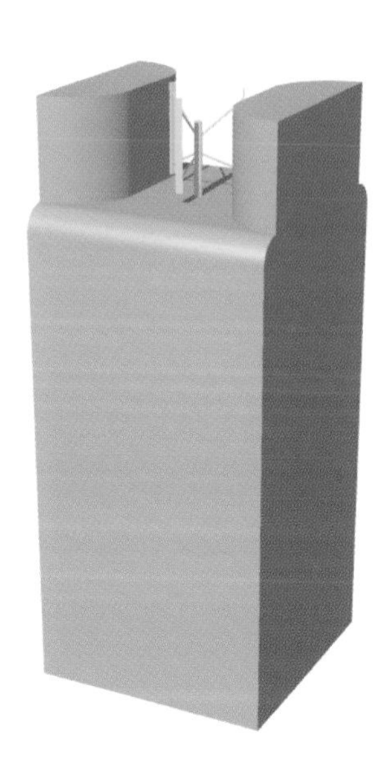

- The building spaces which act as concentrators can be occupied if suitable acoustic buffers are provided.
- Yawing may be an issue, therefore a VAWT may be preferable. A HAWT could be used and it may be possible for some turbines to fix the yaw and allow the blades to actively pitch control in order to make use of wind from both directions.
- The low tower will mean 'fall over distance', tower erection, access for maintenance and vibration will be easier to reconcile.
- A lower tower and adjacent structures will mean the blade flicker, visual impact and noise emissions will have a degree of additional mitigation.

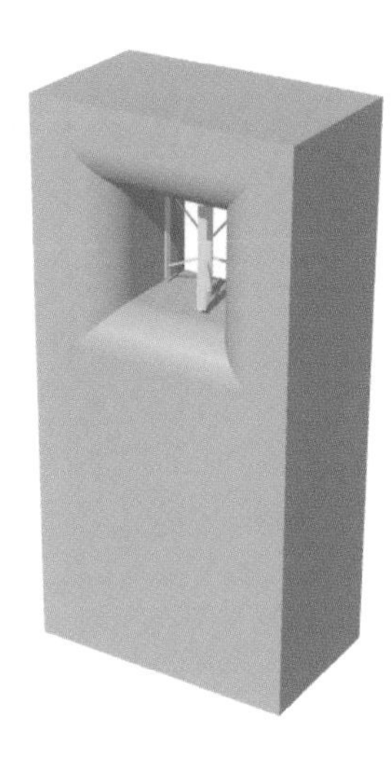

D) Square concentrator within a building façade

This option takes advantage of the higher quality winds at higher altitudes and substantial local acceleration (even if the wind distribution is the same for all directions a 25 per cent energy increase over a free-standing equivalent can be achieved with an increase of 40 per cent for bi-directional winds – see Table 5.2). Although this option requires a loss of lettable space there are a number of examples of large/tall buildings replacing lettable area with a 'feature opening' – e.g. for aesthetics, sky gardens or to relieve wind loading. In this case a feature opening can be used to generate wind energy. In general:

- This form of integration favours buildings with narrower profiles.
- VAWTs may be preferable as their 'swept area' is square.
- The size of the opening size may depend on the available technology (which is limited) in order to avoid more costly bespoke designs. The option of using an array of units can provide some flexibility and may be useful to consider for larger openings.
- The buildings spaces adjacent to the turbines should be acoustically and thermally insulated (and not glazed).
- Safety should be well considered, e.g. for access for maintenance, and in case of blade shedding.

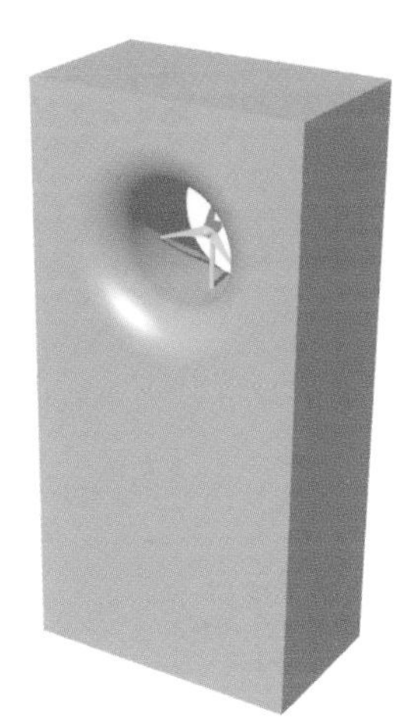

E) Circular concentrator within a building façade

This is similar to the square concentrator with the exception that the shape lends itself to HAWT and energy yields are further increased (for example for a uniform wind a 35 per cent energy increase over a free-standing equivalent can be achieved with an increase of 50 per cent for bi-directional winds – see Table 5.2). In general:

- The size of the opening should be coordinated with the available technology to avoid costly bespoke designs.
- Yawing may be an issue, therefore VAWTs may be preferable. If fixed yaw HAWTs are used active pitch control, available on some of the larger turbines, can be used to make use of wind from both directions.
- Incorporating the cylindrical shroud and the curved façade will be more expensive than a more standard square opening, although the aesthetics character may be more suitable if the form of the development is suited to curved forms.

F) On the side of a building

This option takes some advantage of the higher quality winds at higher altitudes. However, unless the building form is optimized for the local wind character it is likely that the turbines will not perform as well as free-standing equivalents (around 80–90 per cent of the total energy – see Table 5.2). In general:

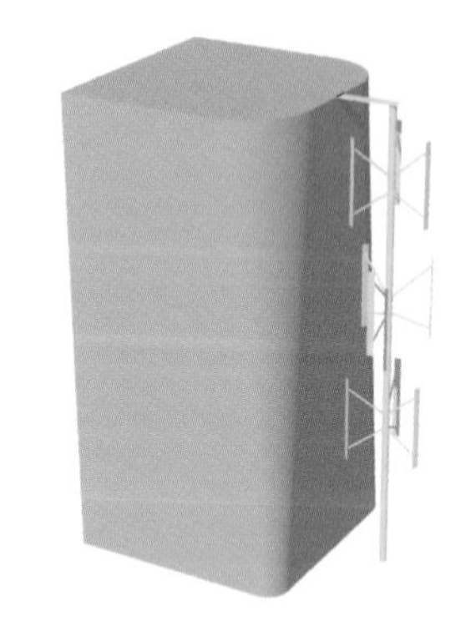

- Yawing may be an issue, therefore VAWTs will usually be preferable.
- The 'swept area' may be maximized by using several turbines.
- The building spaces adjacent to the turbines should be acoustically and thermally insulated (and not glazed).
- Safe and reliable turbines should be used as safety and access for maintenance will be issues, therefore less efficient drag type VAWTs may be preferable.

G) Between multiple building forms

A range of architectural forms are possible when a multi-building development is being considered. Significant local acceleration can be achieved for reasonably basic, non-optimized forms (around 10 per cent extra energy compared to a free-standing equivalent – see Table 5.2). In general:

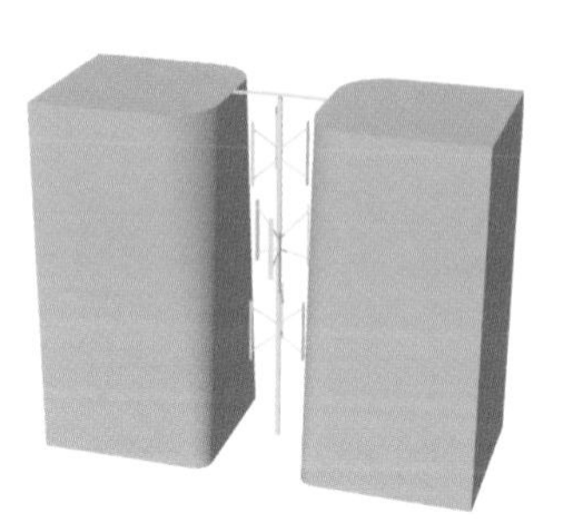

- The orientation/form/shape/separation distance of the buildings will be key variables in the performance of the turbines.
- Yawing may be an issue, therefore VAWTs may be preferable.
- The 'swept area' may be maximized by using several turbines.
- The buildings spaces adjacent to the turbines should be acoustically and thermally insulated (and not glazed).
- Safe and reliable turbines should be used as safety and access for maintenance will be issues, therefore less efficient drag type VAWTs may be preferable.

Other multi-building configurations are considered again later in this section and presented with their energy generation potential.

The idea of using a building to allow turbines to reach greater heights is exemplified by the proposal for the 400m high Dubai International Financial Centre (DIFC) Lighthouse Tower. This has three limited-yaw 29m HAWTs (total peak energy production of 675kW) and a safety screen that could double as windward directional wind vanes. The Lighthouse Tower design is reported to provide energy savings of 65 per cent against standard tower designs, with 10 per cent of the building's energy needs coming from renewable technologies. Of course having turbines at this height will create maintenance issues and crane access considerations will exist not only during construction but after completion of the public realm.

2) WIND DIRECTIONALITY AND BUILDING ORIENTATION

For most urban wind energy proposals it is necessary to determine how the proposed site relates to the directionality of the wind (i.e. which directions contain the most energy on an annual basis). For the building-integrated turbine types B–G given in the previous section, the appropriateness of the design will depend strongly on the directionality of the wind.

Table 5.1 develops the original wind classifications proposed in Project WEB. This was based around historical wind data and analysis techniques contained in the European Wind Atlas (EWA) – a product of an EC research project. Statistical data in the EWA is divided into twelve 30° sectors from which the wind is incident.

Table 5.1 Wind directionality classification for building-integrated turbines

Classification	Criterion	Suitable generic types
Weakly uni-directional	>60 per cent of the annual mean power density in the wind comes from any given 150° wind direction sector	A, B, G
Uni-directional	>75 per cent of the annual mean power density in the wind comes from any given 150° wind direction sector	A, B, C, D, E, F, G
Bi-directional	>95 per cent of the annual mean power density in the wind comes from two opposite 150° wind direction sectors (and each sector produces at least one-third of the sum of the two sectors – i.e. 25 per cent of the total).	A, B, C, D, F, G

Several urban sites possess wind regimes that are either uni-directional (e.g. Lisbon, Athens, Toulouse, Munich and Stuttgart) or bi-directional (e.g. Lyon, Seville and Frankfurt). A significant number of sites can, however, be considered as weakly uni-directional (e.g. Dublin, Paris, Essen and Birmingham).

Table 5.2 provides data on the theoretical annual energy yield increase from an investigation carried out by BDSP Partnership on the seven fundamental building integration types. The energy yields are given in relation to a free-standing equivalent turbine at the same height for the three separate wind types as defined in Table 5.1 plus the theoretical uniform condition representing the same wind energy from each direction. The results shows that for these basic, non-optimized building forms a significant energy increase is possible for most cases even in uniform winds (which have the same wind energy content in each of the 12 wind sectors).

Table 5.2 Total energy from different building-integrated turbines (relative to a free-standing equivalent at the same height) for different wind types

Geometry type	Overall power production change for various wind type			
	Uniform	Weakly unidirectional	Strongly unidirectional	Bi-directional
A	1.12	1.12	1.12	1.12
B	1.15	1.15	1.15	1.16
C	1.03	1.07	1.10	1.16
D	1.24	1.27	1.31	1.38
E	1.35	1.39	1.43	1.51
F	0.78	0.81	0.84	0.88
G	0.99	1.03	1.07	1.13

These results only factor energy increase from local wind concentration effects. The energy yields will increase further (over standard ground-based installations) if the turbines can be elevated into higher quality winds using the building form.

It should be noted that the data presented in Table 5.2 is presented for indicative purposes only. The results are will vary in both the positive and negative direction depending on the particular real-life project being considered. Approaches to predicting energy yields are explored in the next section.

Once the historical wind data for a certain locality has been analysed the design of a particular wind energy development should be tailored to the local wind conditions. For example, a site with an omni-directional or uniform wind, which will be common in many urban areas, would call for designs able to make the most of the available resources.

Figure 5.3 Omni-directional concept tower (Project WEB – ibk2 University of Stuttgart)

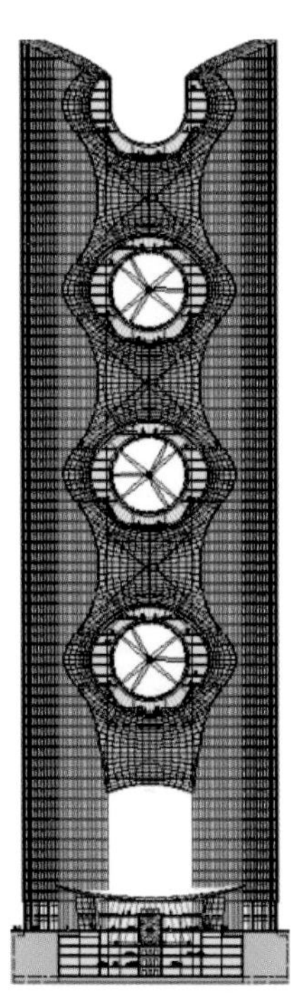

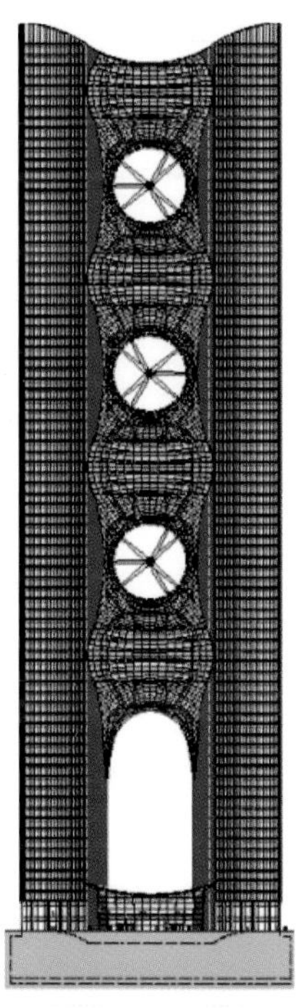

An example of an omni-directional wind energy tower is shown in Figure 5.3. It is based on four 360m tall towers containing office spaces, centred around six turbines stacked one on top of the other. The turbines are suspended in ducted holes spanning multiple storeys, which connect the four towers (which form their inlets and outlets). There are two sets of ducted holes running at 90° to one another for alternating turbines, i.e. one set of ducts running north–south and one set east–west to capture all these wind directions.

The omni-directional tower can be designed with either HAWTs (as shown in Figure 5.3) or VAWTs and will harvest the incoming wind from all directions.

Wind tunnel tests and computational fluid dynamics (CFD) studies found that optimum performance occurred when the wind was at 45° to the towers of the building. The streamlines would curve and straighten through the ducts so that both banks of turbines would be spinning. Even with the wind at right angles, one set of turbines would always be operating. The use of criss-crossing ducted holes creates unusual voids above and below the turbines in the centre of the building. This could provide some design issues for architectural integration, although there is scope for the creation of interesting design spaces. Many fundamental design details would, however, need to be resolved on the buildings in terms of the structural, mechanical, electrical and acoustic design.

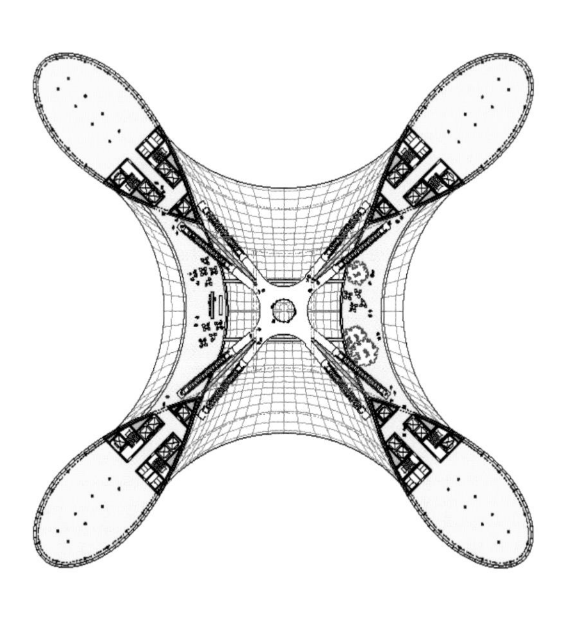

3) PREDICTING ENERGY YIELDS FROM TURBINES INTEGRATED INTO SHROUDS WITHIN TALL BUILDINGS

If urban wind turbines are to be successful and be perceived as a success (and inspiring) by the public, one thing is certain – the blades have to turn. Therefore methods for reliably estimating the energy that could be generated by a particular installation are desirable. With building-integrated wind turbines, these methods will not only be able to predict the energy yields but provide a means to improve the building form in the design stage to maximize energy production. Any performance prediction method will ideally need to be reasonably accurate without incurring prohibitive costs and time penalties.

The most reliable method is large-scale physical testing. This type of testing was carried out extensively during Project WEB. The prototype had a number of noteworthy features:

- It consisted of two independent symmetrical towers (7m high x 1.5m wide x 2.3m deep) based on the aerodynamic 'boomerang' profile.
- The tops of the aluminium-clad towers were sloped for aesthetic rather than aerodynamic reasons.
- The building was mounted on a base frame whose wheels sit on a circular rail. This allowed the building to be easily orientated into the wind in order to efficiently capture experimental data.
- The turbines sat on their own tower (passing through the centre of the base frame), with their rotors at the narrowest point between the towers at a hub height of 4.5m.
- Aerodynamic infills linking the towers were added after initial testing to produce a ducted hole in which the turbines sit. They were designed to be aerodynamic, aesthetically pleasing and practical to construct, rather than just to produce optimal wind enhancement.

Two electrical systems were also designed:

- a control circuit for the wind turbines, whose electrical power is stored in a battery with a load taking charge from the battery;
- an instrumentation system for automatically measuring and logging climatic data (e.g. wind speed and direction from two site anemometers) and data from the wind turbine system at two-second intervals via a personal computer.

The results for this non-optimized design were encouraging. Placing the wind turbines inside the building/concentrator produced considerably more power compared to when they were conventionally mounted at the same height on an open site. Specifically:

- The peak value of the coefficient of performance 'Cp' was doubled by placing the turbine between the aerodynamic towers (i.e. a doubling of peak power).
- The integrated turbine produced enhanced power output for a wide range of incident wind angles (± 75°) onto the building.
- The optimum power enhancement occurred when the wind was incident at around 30° onto the building/turbine rather than at a wind angle of 0° ('zero yaw').
- The addition of the infills further enhanced concentrator performance (particularly at acute wind angles).
- The concentrator increased the 'effective wind speed' seen by the turbine by a significant 1m/s.

Even when the wind was at an angle of 90° (impinging onto the sides of the towers), the HAWT still produced substantial power (50–90 per cent of that of a stand-alone turbine for the same wind speed). This appeared to be due to complex flow phenomena, i.e. flow remaining attached between the towers while the wind direction fluctuates rapidly. Similar trends were seen when a VAWT was integrated into the prototype building. The performance graphs, with and without the accelerating infills, are reproduced in Figure 5.5.

Figure 5.4 Prototype building designed for the integration of single HAWT or VAWT wind turbines (with or without 'infills') from Project WEB (BDSP Partnership, MECAL, Xkwadraat (NL), CRLC RAL)

Performance with infills

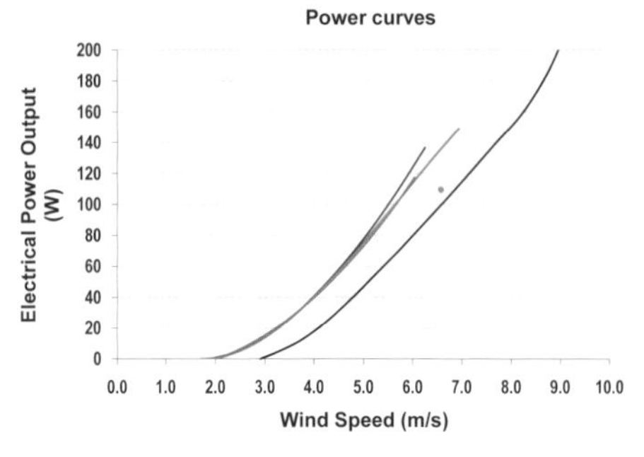

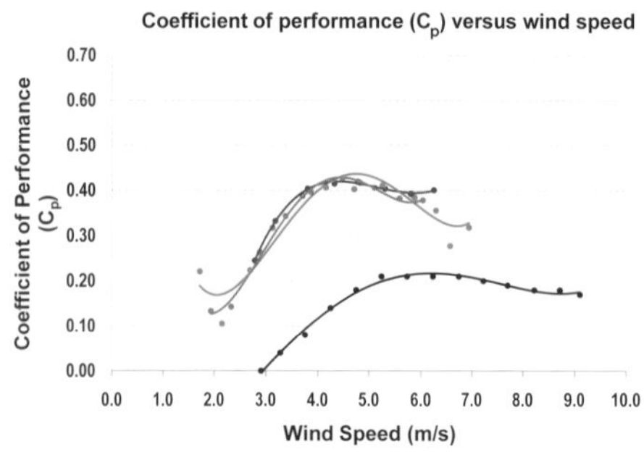

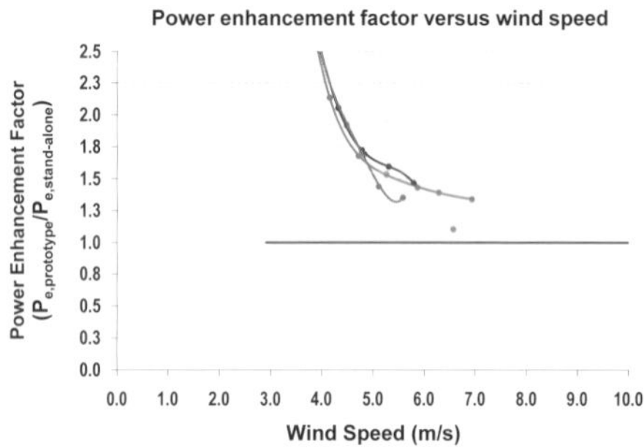

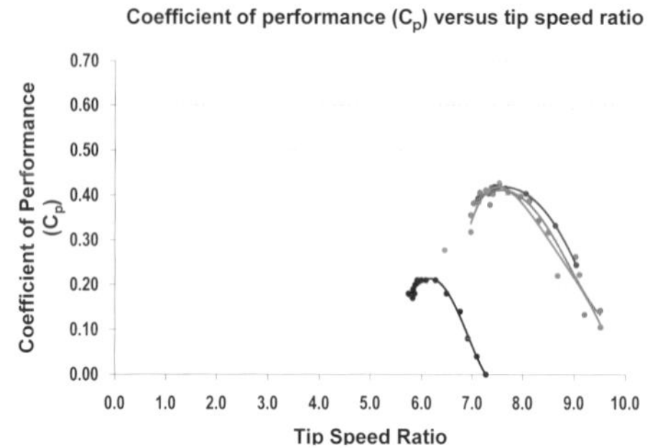

Performance without infills

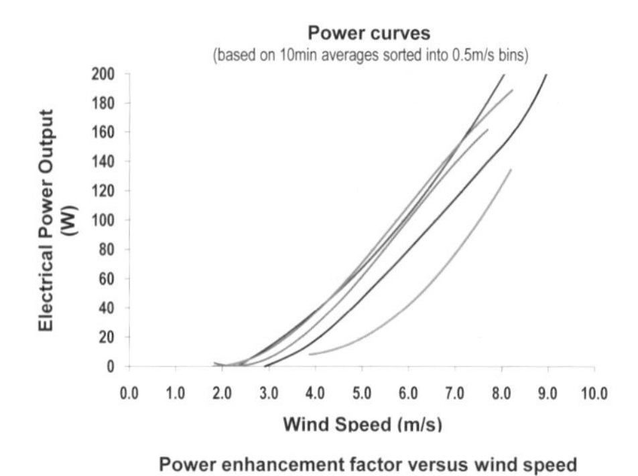

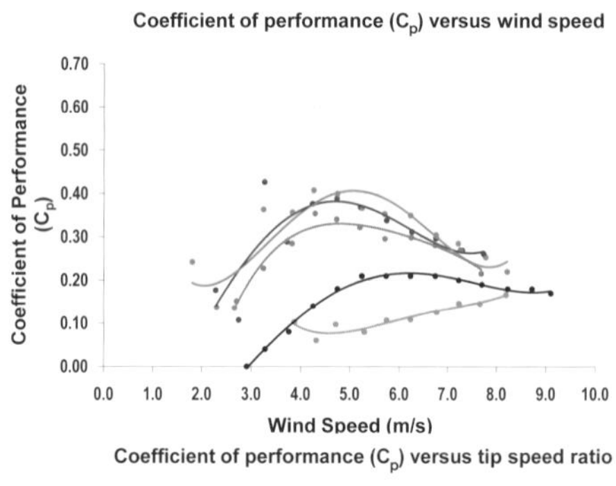

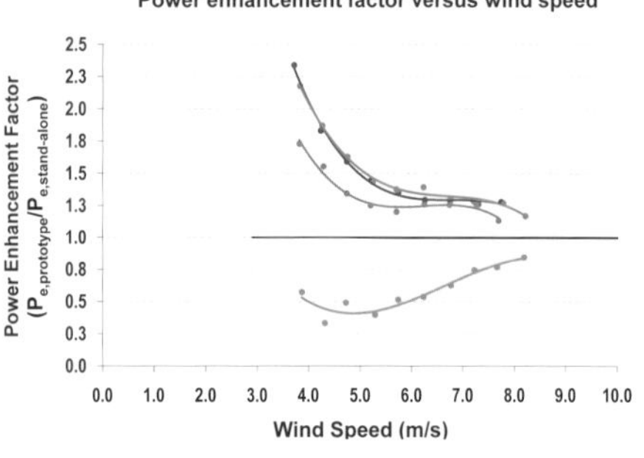

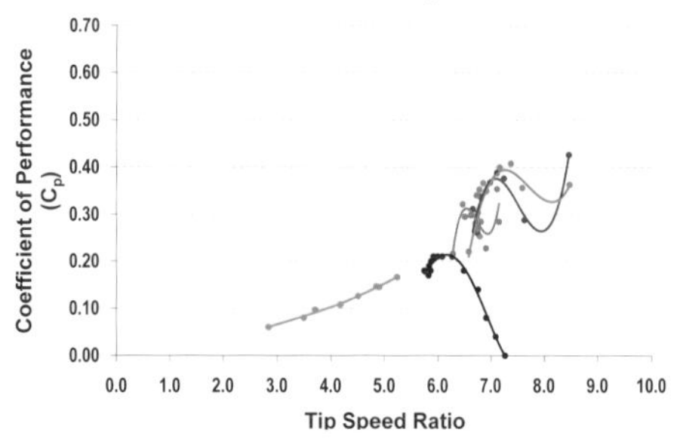

Sector 90
Sector 60
Sector 30
Sector 0
HAWT Stand-Alone

Note : all results are based on 10 minute averages sorted into 0.5m/s bins.

Figure 5.5 Energy performance of the integrated turbines within the Project WEB prototype building – with and without concentrating 'infills' between the aerodynamic towers (BDSP Partnership)

Figure 5.6 Prototype small-scale wind tunnel architectural models designed for the integration of single or multiple wind turbines (Project WEB – ibk2 University of Stuttgart)

These experimental results have been sorted into 30° sectors to highlight the dependence of performance on the angle of incidence of the wind onto the prototype. 'Sector 0' includes angles of incidence of ±15°centred on 0°; 'Sector 30' includes angles of incidence of (±15°- ±45°), i.e. two symmetric 30° sectors centred on ±30°; and so on.

It was concluded that with this type of building-integrated turbine it should be possible to increase energy yields by at least 50 per cent for most urban sites provided that the building shape is suitably aerodynamic. This is in line with the data from the theoretical studies (derived from CFD simulations) presented in Table 5.2 for turbine integration type E.

It should be noted that Project WEB was a research project partly funded by the EC and typical commercial projects will encounter obvious limits in terms of large-scale physical testing due to associated costs and time required to build these types of structures. Despite the reliability of the results for this method, additional drawbacks exist. For example, it can be difficult to model/represent the influence of surrounding buildings when testing a real-life proposed development (unless the tests can be carried out on the actual completed site). Also, optimization can be costly as a degree of rebuilding is required. Furthermore, replicating the same test conditions can be difficult due to the unpredictability of the wind on any given test day.

One method which allows the surrounding buildings to be taken into account is small-scale physical testing (e.g. 1:200 or 1:400 scales) in wind tunnels. Time scales, costs and controllability of test conditions are considerably improved compared to large-scale physical testing. However, although this is the accepted standard for assessing wind loadings on building façades, extending this technique to this area can have drawbacks. For example, shear and turbulence do not scale down particularly well and it can be difficult to pick up on the finer details of geometry optimization and flow measurement at these small dimensions, e.g. a 10m turbine will only be 2.5cm in a 1:400 model. Nevertheless, even though it is not possible to create a realistic model of a fully functioning turbine at small scales, wind tunnels have been used to produce useful results.

Calibrated gauze discs of known resistances can be used to represent wind turbines using a validated experimental technique. By measuring the streamwise force acting on the disc (using strain gauges) it is possible to estimate the power that a wind turbine would produce.

Figure 5.7 Wind tunnel and dye tracing techniques for analysing building-integrated wind turbines (Project WEB – Department of Aeronautics, Imperial College, London)

Desktop modelling, via CFD, offers the mode of entry into energy yield prediction in terms of time, capital expenditure and quality of output. However, as with most technology, the accuracy of the results depends heavily on the user. In general, results can be considered acceptable to identify key trends if the models use:

- correct inlet/turbulence profiles;
- high order solutions;
- large enough fluid domains;
- suitable turbulence models;
- good convergence criteria;
- meshes that produce grid-independent results.

This approach to energy prediction can be very quick in terms of man-hours to set up a model once a user has gained enough experience. The simulation run time and the maximum mesh resolution depend on the available computer hardware. Simulations can be set up and run within design team timescales if parallel computing techniques are used – i.e. where dedicated computing clusters or computers linked across a network are available.

The Castle House residential tower, shown in Figure 5.8, was simulated by BDSP Partnership as part of the pre-planning assessment stage (planning permission was granted in 2006 with construction beginning in 2008). The aim was to provide some assessment of energy yields from the three 9m diameter turbines proposed to be located within the three dedicated shrouds at the top of the building. The flow domain, which included the local existing buildings, was 'discretized' into 4 million cells and a high order advection scheme was used to capture the accelerating winds within the shrouds. In each

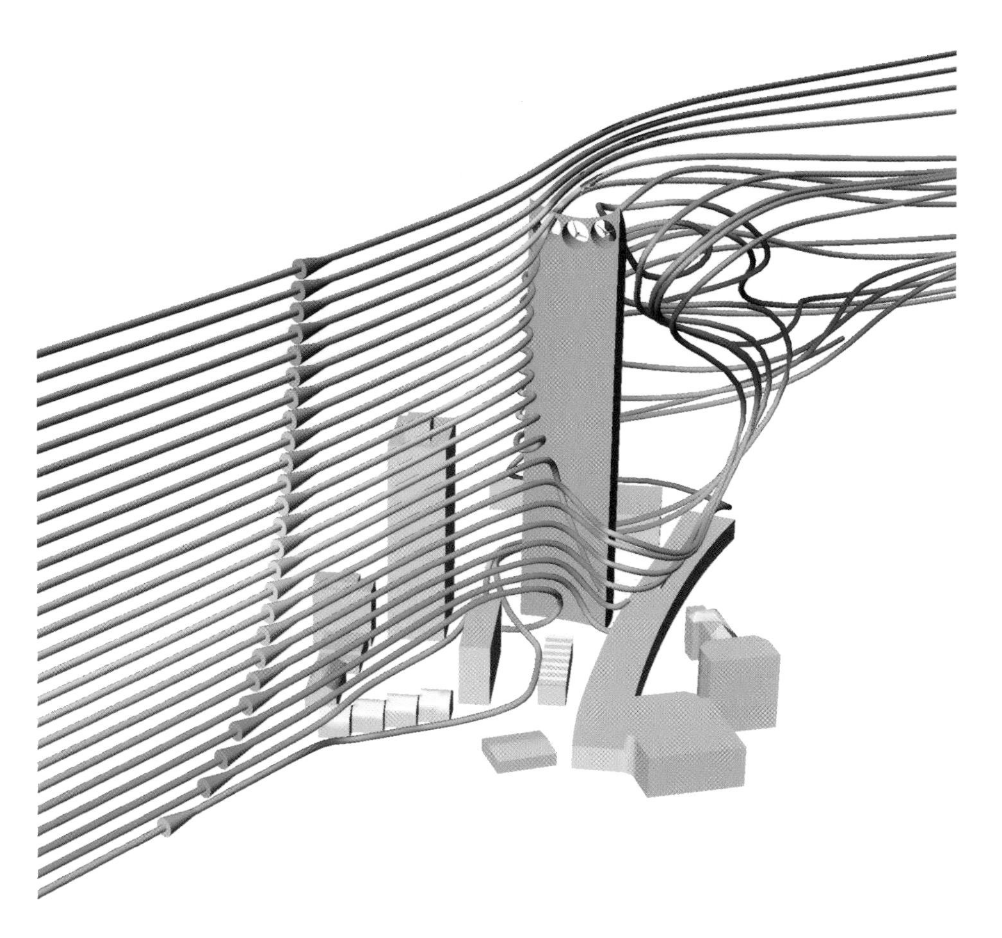

Figure 5.8 CFD model of Castle House, London, revealing the urban wind profile, local downdraughts and swirl character of the wake (BDSP Partnership)

case the steady state solution was converged to very low residual errors. The shear stress transport (SST) model was used to represent turbulence in the flow.

The results in Figure 5.8 show streamlines, coloured by velocity, which originate from the prevailing wind direction and reveal the upstream urban wind profile. At the base of the tower the downdraughts can be seen and the swirling low-speed character of the wake can also be appreciated. As expected, the velocities are greater at greater heights.

The simulation can be automated to run for a variety of different wind directions (in, say, 20° or 30° increments). In each case, the energy content of the air flowing through the swept area of the blades within each shroud can be determined. This can be repeated relatively easily for different geometrical configuration (e.g. for differing shroud designs) and compared directly to free-standing equivalents at the same height. It is then possible to assign relative weightings to the results for each direction corresponding to the total energy content of the given wind direction using climatic data (e.g. for Castle House the wind direction 210° from north would have the highest weight as the winds from this direction are more frequent and strongest). Several geometrical configurations of the top of the building were investigated:

1. the original concept with 9m diameter shrouds;
2. the same shrouds with a 1m radius filleted edge and rounded rear;
3. 7m shroud with a 2m radius filleted edge and an altered canopy.

The results for these three options are given in Figure 5.9. These show a 3D axonometric view of the top of the tower (with velocity contours and vectors plotted on a horizontal plane through the centre of the shrouds) and a section through the centre of the middle shroud for each case.

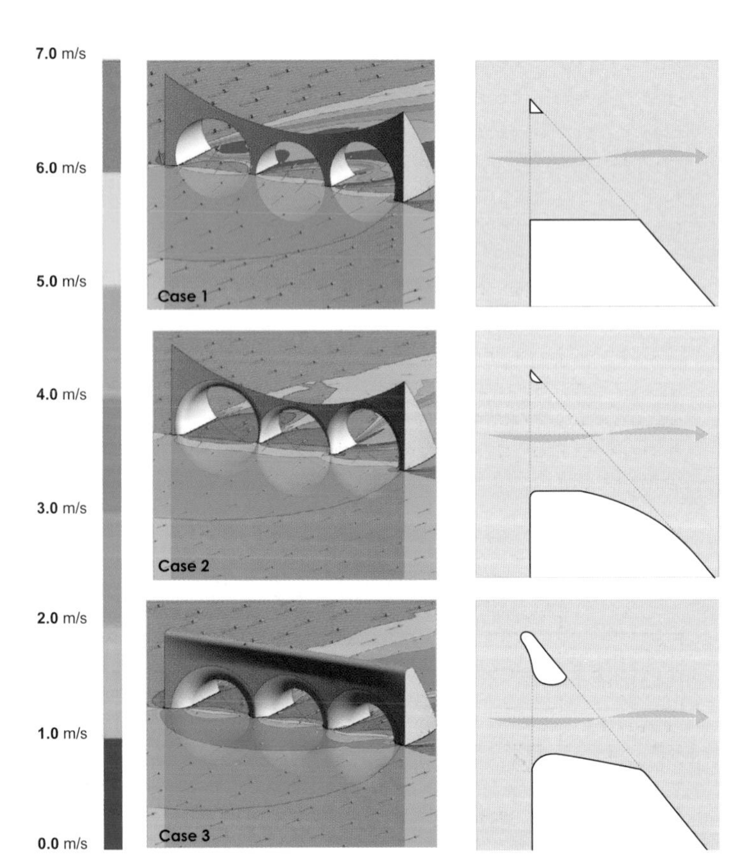

Figure 5.9 CFD models for three Castle House shroud forms - axonometric and section view (BDSP Partnership)

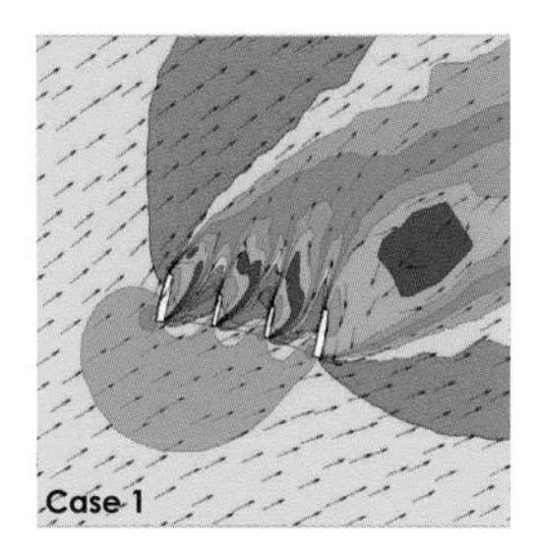

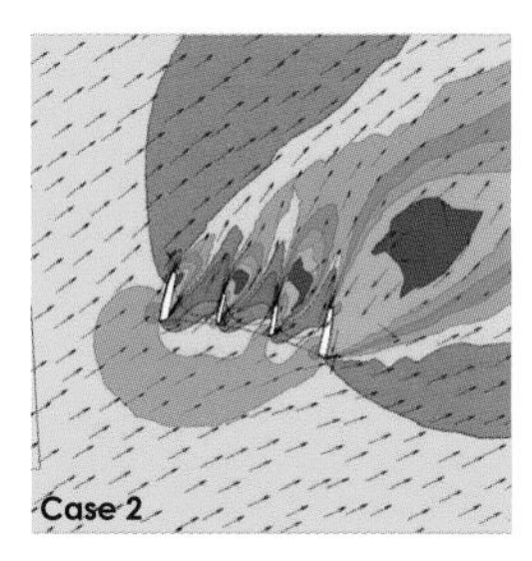

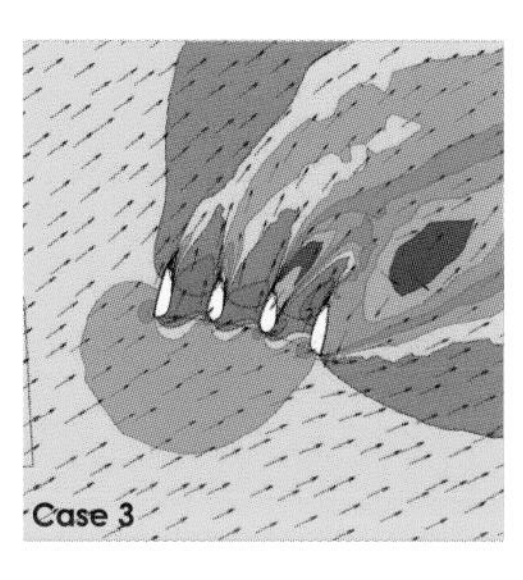

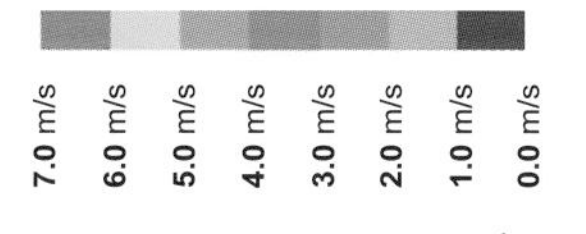

Figure 5.10 CFD results for each of the three Castle House cases in a wind 240° from north – plan view (BDSP Partnership)

The velocity contours and vectors plotted on the horizontal plane through the centre of the shrouds are presented again in 'plan' for each case (Figure 5.10). These results are for a wind direction 240° from north. This corresponds to a wind 45° from the 'ideal' wind which would be normal to the fixed plane of the turbines. Moving from Case 1 to Case 3, an increase in wind speed through the shroud is noticeable. This is primarily due to a reduction in flow separation resulting from rounding off the sharp edges. In Case 3 the wind is able to be redirected by the shroud through 45° and will tend to meet the blades at the correct angle, which is a very important consideration.

When calculating the energy increase associated with wind acceleration, the most common pitfall relates to the assumption that the energy in the wind corresponds solely to the wind speed, i.e. kinetic energy component. This assumption follows on from the well-known wind energy formula used for wind turbines in the free stream. However, when there are significant local obstacles and pressure drops, the energy should be evaluated on a 'total pressure' basis which is the sum of the kinetic energy *and* the local static pressure. It can be common to assume that a significant amount of energy is 'given' to the wind when accelerated but the increase in velocity results in a decrease in pressure energy (from Bernoulli's equation). There are of course energy benefits in accelerating winds but they are proportional to the increase in mass flow rate and not proportional to the cube of the wind speed.

The results of the Castle House simulations, given in Table 5.3, are therefore presented in terms of mass flow through the swept area of the three shrouded turbines for each case. These mass flow values are normalized with respect to a free-standing equivalent. For example, a value of 0.5 corresponds to half the energy generated by a free-standing equivalent turbine at the same height.

Table 5.3 Mass flow through the Castle House turbines, for varying shroud designs, normalized with respect to 9m diameter free-standing equivalents (for 4 wind directions)

	Wind direction (deg from North)			
	180	210	240	30
Case 1	0.8	0.9	0.7	1.3
Case 2	1.1	1.4	1.0	1.4
Case 3	1.3	1.3	1.1	1.0
Case 3*	2.2*	2.2*	1.8*	1.7*

* normalized with respect to a 7m diameter equivalent turbine

As the results show, the original configuration generates less than a free-standing equivalent for most wind directions. Significant improvements are made by reducing the extent of the flow separation (Case 2). However, switching to a 7m diameter shroud, with a 2m radius fillet, improves the energy performance despite the decrease in swept area of the turbine.

Historical hourly wind data from the CIBSE test reference year for London were used to estimate the annual energy generated.[2] Considering only the wind from a south-westerly sector spanning 130° (i.e. ±65° from the normal to the plane of the turbines), Case 3 is predicted to produce more energy than equivalent 7m diameter free-standing turbines (at the same height) which can take energy from the full 360°.

Even without taking all 12 wind directionS into account, confidence in the design can be drawn from this type of wind energy integration proposal, especially as the design is able to produce more than double the energy output for the key wind directions. It should be noted that at the top of the building, some 150m above ground level, the turbines will be exposed to relatively high wind speeds. However, these wind speeds may, depending on the location, remain below wind speeds in neighbouring open areas at much lower heights.

This method has proved viable from a man-hour/cost point of view over and above physical testing. With regard to the accuracy, the standard k-epsilon method of predicting turbulence, although reliable in the free stream, can produce errors close to obstacles including over-predicting pressures. Other models can be used, such as the low Reynolds number k-epsilon. This resolves the boundary layer more accurately if care is taken with the boundary layer mesh and is not much more CPU intensive. The SST turbulence model used in these cases is a hybrid statistical model which also resolves the boundary layers when needed. Ideally the transient nature of the wind would be modelled, e.g. to capture vortex shedding effects. Although this can be done for simple geometries using simple two-equation models like the k-epsilon model, it is perhaps better approached using large eddy simulation (LES) or detached eddy simulations (DES) models.

Despite the approximation of the isotropic statistical representation of turbulence provided by the two-equation model it should be noted that good agreement was found between CFD and large-scale physical testing within Project WEB.

BOX 5.1

POTENTIAL PITFALLS WHEN ACCELERATING WINDS

The pitfalls of deliberately accelerating winds relating to pedestrian comfort, position of ventilation opening and the opening of windows have already been mentioned. However, there is another potential pitfall that should be avoided when accelerating wind. This relates to additional heat (or 'coolth') loss which can arise when accelerating the wind.

The heating and cooling loads increase when wind speeds are deliberated, or otherwise, accelerated near building façades. Any additional building energy loads incurred as a result of artificial wind acceleration will of course act counter to the objectives of installing turbines.

The impact of the energy demand on the building will depend on the extent of the wind acceleration (m/s), the area (m^2) of the affected façade and the insulation of the affected façade (W/m^2K). In order to give a feel for the orders of magnitude of losses, two façade constructions are now briefly considered. In these two cases 10 per cent of the building envelope is assumed to be affected by wind acceleration and the external resistance 'R_o' is assumed to decrease from $0.04m^2K/W$ for a typical 5m/s wind condition to $0.02\ m^2K/W$ for an accelerated wind condition where the wind speeds are 9m/s on average:

1 A high insulation case – e.g. a wall with a U-Value of $0.25W/m^2K$ would give an overall load increase of ~0.05 per cent.

2 A low insulation case – e.g. a double-glazed façade with a U-Value of $2.0W/m^2K$ would give an overall load increase of ~0.4 per cent.

The values of the resistances used for the calculation are derived from CIBSE (R_i and R_w relate to the resistance to heat flow from the internal boundary layer and wall respectively).[3] The values of the three resistances which contribute to the U-value will vary on a case-by-case basis.

$$U = \frac{1}{R_i + R_w + R_o}$$

This effect, i.e. the amount of extra heat or 'coolth' required to compensate for the deliberate wind acceleration, will be more pronounced in regions that have either a very cold climate (where the heating load is high) or a very warm climate (where the cooling load is high).

If the additional heat and cooling loads are calculated to be a significant proportion of (or more than) the energy generated from a proposed wind energy installation the project should be re-evaluated or redesigned e.g. to provide more thermal insulation in the appropriate areas.

4) ENVIRONMENTAL IMPACTS, BUILDING DESIGN AND PLANNING

For a given building-integrated wind turbine project, the planners will have to be assured that the environmental impacts are well within acceptable limits. Psychological acceptance of the need for wind energy and renewable energy will clearly be important in the decision-making process of planners or stakeholders, and decisions made one way or the other may reflect cultural differences between countries. However, developments that may harm the local environment by imposing unnecessary additional safety risks, visual intrusiveness and noise problems, will not proceed in any region.

Visual impacts

Generally, the visual impacts of turbines are one of the first aspects to be considered. This may be the result of cases highlighted by the popular media where wind farm developments in rural area have been challenged as they can be seen to be damaging to the intrinsic beauty of the natural areas with knock-on effects on tourism and property values. The aesthetics of urban landscape could also be damaged if a turbine is erected without careful consideration. Large turbines mounted at roof level may be visible from many different vantage points over several kilometres. However, as man-made objects, wind turbines may be considered more visually suited to the urban environment.

The visual and aesthetic appropriateness of a turbine is linked to the perceived 'story' behind the development. For example wind energy may be thought of as more appropriate to urban areas as it provides energy direct to the end user.

Acceptance by planners and neighbours may be low if wind energy is seen as a poorly considered afterthought or as last minute 'greenwashing', e.g. a wind turbine stuck on top of an otherwise seemingly standard building. However, when the building is clearly designed to work with wind energy, the idea associated with visual impacts may be completely turned around especially if the development is linked with other visible green design features. In these cases, instead of minimizing the visual nature of the building, a turbine which is an integral part of a building will have its own inherent aesthetic appeal. In these situations the development as a whole (turbine included) can become a showcase for innovative design, creativity, prosperity and care for the future of the planet.

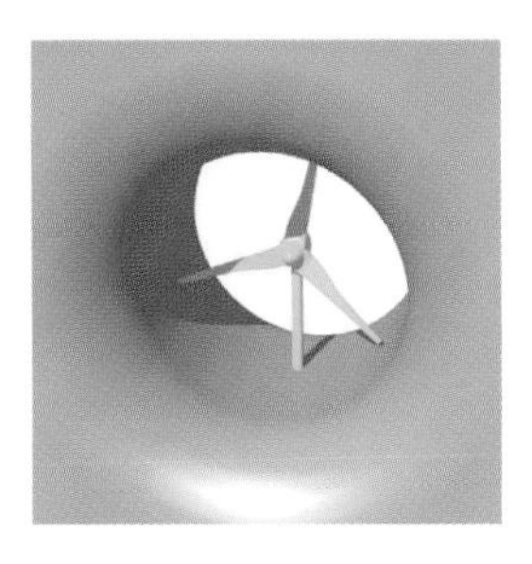

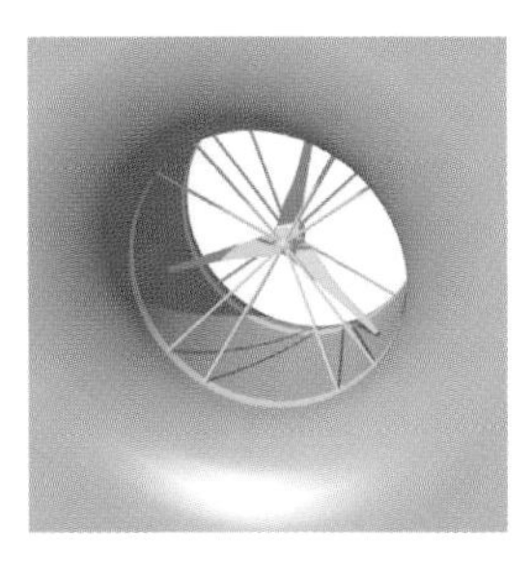

Figure 5.11 Three types of turbine support: off-the-shelf tower, aerodynamic bridging and suspension rods (Project WEB - MECAL, BDSP Partnership)

Safety and turbine suspension

Safety and reliability of a building-integrated turbine is a key issue and the means to link the turbine to the building – i.e. the means of suspension – is an important area to consider.

Standard, off-the-shelf, tried and tested components, such as standard towers and nacelles, will be favoured over more expensive bespoke elements. For many building-integrated wind turbines these standard components can be used.

In certain circumstances modifications to existing components can be relatively simple, for example when securing a VAWT from the top as well as the bottom. In some cases more sophisticated means of suspending wind turbines may be required, e.g. bridging or suspension rods (which can also form the basis of protective screening).

All forms of turbine support will tend to reduce the flow of air and therefore the associated energy content available to harvest. They will also be subject to variable aerodynamic and mechanical forces which cause vibration. The reduction in air flow can be mitigated by producing an aerodynamically shaped support, as shown in the bridging example above. The vibration issue is more complex since the building and turbine will interact and impact on one another. These induced vibrations will affect not only the turbine suspension, but also the blades and nacelle of the wind turbine itself and the points at which the turbine and suspension are supported. The example of the turbine suspended by several rods attempts to reduce aerodynamically induced vibration by having the suspension angled away from the fastest moving part of the blades (the tips).

Computer tools based on finite element methods and CFD can be used to examine vibration problems, although the complexity should not be underestimated. For example, resonance problems can occur at different rotational frequencies and there may be a broad range of these operational frequencies, for example if the turbine is a variable speed machine or when a wind turbine starts up or stops. Specialist consultants should be brought on board to deal with these complex issues.

As the 'bridging' and 'rod-based' types of suspension are not commonplace, design criteria may need to be developed based on existing wind and construction industry standards. These standards will vary between countries and the integrity of the designs may have to be certified.

In general the building and turbine suspension must perform several basic functions:

- The turbine suspension must be capable of supporting the weight of the turbine, and the building in turn must be capable of supporting both.
- Acceptable deformations of the building and suspension must be achieved under all design (load) conditions.
- The turbine suspension must not fail due to material fatigue caused by extreme wind loads and cyclic loading induced by fluctuating wind conditions and rotation of the turbine over its operational lifetime.
- Damage to the building structure, damage to plant and equipment and nuisance for the building occupants due to vibrations induced by the rotation of the turbine must be minimized.

Figure 5.12 Suspension cables on an architectural model (Project WEB – ibk2 University of Stuttgart)

Due to the close proximity of people and façades, 'fail to safe' suspension should be designed wherever possible.

During Project WEB the structural optimization of the turbine suspension showed that it should be possible to create an aesthetic, streamlined, aerodynamic design. The final turbine suspension weighed less than two-thirds of a conventional turbine tower. Also noteworthy, is the assessment of the coincidence of blade pass frequencies and the natural frequencies of a conventional square-edged building which suggested that resonance would be a problem during normal turbine operation. Vibration control at source, i.e. within the turbine suspension, was deemed necessary.

Suspension of turbines will inevitably have significant associated impacts on the structural systems used within the building, which will require stiffening and (passive or active) vibration control measures to cope with both the static and dynamic loads induced by the weight and rotation of the turbine(s).

Safety, in certain circumstances, can be given an elevated status. One way to enhance safety would be to re-design the turbines themselves. For example, a chain can be placed inside each blade to keep the blade together in the event of a failure. While 'invisible' safety improvements are undoubtedly important, there may be a need to provide 'visible' safety devices which crucially enhance both safety (of people and property) and the public perception of safety. In its most basic form, this may be a safety cage placed around a wind turbine as shown in Figure 5.13.

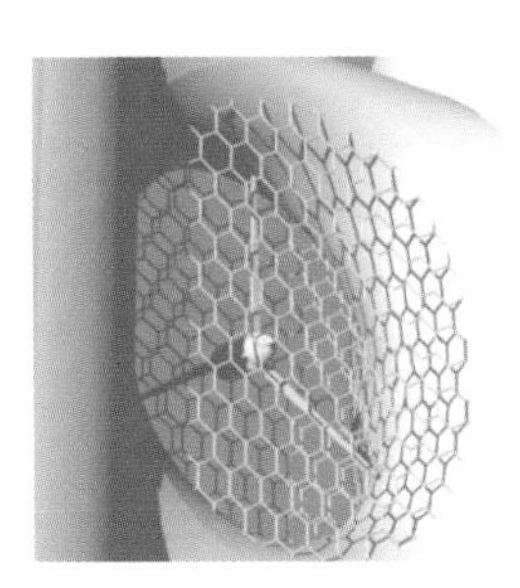

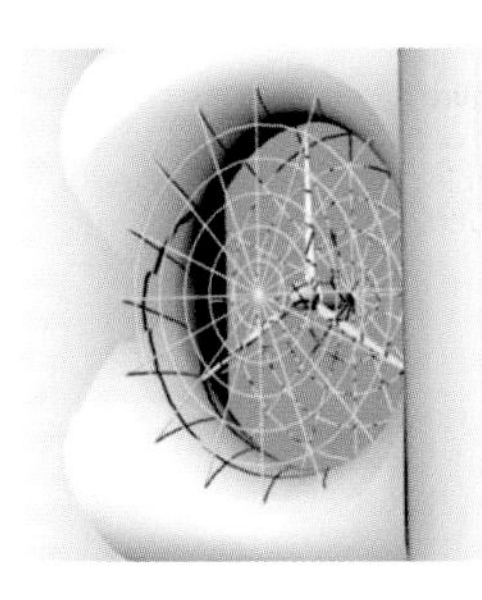

Figure 5.13 'Visible' honeycomb and radial mesh safety devices (Project WEB – MECAL, BDSP Partnership)

Performance requirements and design options for external safety devices have been identified (e.g. position, design, materials, level of automation etc.). For example, a safety device would ideally:

- enhance safety and the perception of safety;
- be environmentally friendly, for example reducing the likelihood of bird kills;
- be capable of withstanding exposure to climatic elements over the design life;
- be structurally sound and capable of absorbing impact energies;
- reduce average wind speeds through a ducted hole as little as possible;
- minimize turbulence generation;
- be aesthetically attractive;
- reduce shadow and light flickering from the blades;
- be an acoustic absorber rather than propagator of noise;
- be easily transportable and maintainable;
- reduce electromagnetic interference;
- be cost-effective.

Conceptual designs for a prototype safety device to be fitted to a real building where turbines are suspended within ducted holes have been carried out. One of the best options was found to be a radial mesh safety device positioned both in front of and behind a wind turbine and attached to the building.

Noise Emission of building-integrated turbines

Noise as an environmental impact has been discussed in some depth in Part 3. However, the noise emission and propagation of building integrated turbines can involve two additional complicating factors:

- Complex propagation due to reflections from roofs and façades;
- Additional noise source such as vibration amplification of structures.

Noise can penetrate into a building either directly via airborne propagation or indirectly via transmission by vibrations of the external envelope. In order to quantify the noise emission and propagation for a given scenario computer simulation techniques can be used.

Sophisticated analytical and computational techniques for predicting noise emission from wind turbines are comparatively recent phenomena and not yet widely used. These may be based on the broad characteristics of the turbine (e.g. rotational speed, hub height) or consider each possible noise mechanism in detail.

In the noise prediction models widely used in planning submissions for wind farms, the turbines are modelled very simply as point noise sources located at hub height. They are scaled by sound power levels (SWLs) measured by the manufacturer for the particular wind turbine (quoted for a given rotational speed and hub height).

These models focus on the sound level that would be perceived by receivers at various locations in the far field (i.e. a significant distance away from the turbines), which is normally the prime concern in planning enquiries for wind farms. The sound perceived by an observer is characterized by the sound pressure level – a function of the distance between the source and receiver.

The models normally assume that the source radiates equally in all directions. Assuming simple spherical spreading, i.e. that the sound pressure level (SPL) varies according to the inverse of the square of the distance between the source and receiver. This would mean that the sound pressure level is reduced by 6dB for each doubling of distance.

In reality, there are several other complex phenomena such as atmospheric absorption, reflection from the ground and wind speed and direction (all dependent on the frequency and distance between source and observer), which have to be taken into account to produce accurate predictions. To make matters more complex, these phenomena will vary on both a spatial and time-dependent basis. Consequently a variety of 'noise propagation' models have been developed.

Simulating noise emission and propagation from a building integrated turbine, or even a stand-alone turbine mounted in an urban setting, must primarily take into account the significant influence of the presence of the building in the near-field, particularly if mounted within a ducted hole. Models can also be set up to allow for:

- noise reflection from other highly acoustically reflective surfaces such concrete walkways and tarmac roads;

- noise emissions from the blades continually passing the building and turbine suspension;
- noise emissions via flow-induced vibration due to complex aeroacoustic phenomena caused by wind flow around the building.

Simulation of noise emission and propagation from a prototype

In order to make an initial assessment of noise emission and propagation from a building-integrated turbine, a prototype design based on 100m tall towers and a 30m diameter HAWT suspended over 60m above ground level has been analysed. Specialist building acoustics software was used in order to account for the most important propagation factors, including frequency-dependent behaviour and diffraction around obstacles.

Eight equidistant point sources were placed on the circle swept out by the tips of the turbine blades (termed a 'disc source'). This allows for a more precise representation of the interaction between the tips of the turbine blades and the inner surfaces of the building towers/infills, which is expected to be the dominant noise emission phenomenon.

The eight point emitters forming the disc source were scaled to produce the same total SWL as the single point source model and based on manufacturer's sound power level data for a particular turbine (in this case 95dBA for reference conditions). The effect of wind speed and direction was not taken into account as the near-field perceptions of sound are not greatly influenced by urban wind patterns.

Studies were carried out to give initial predictions of the external sound pressure levels for a HAWT mounted between aerodynamic ('boomerang' profile) towers linked by infills surrounding the turbine.

Subsequent studies were carried out to investigate the choice of materials used for the towers and infills on external and internal receivers, and the effects of noise emissions on surrounding buildings (for an example site).

Virtual planes were placed within the computer models to predict sound power levels at receiver points and then compared to urban noise regulations applied in a number of EU countries, as would be the case in a real planning enquiry.

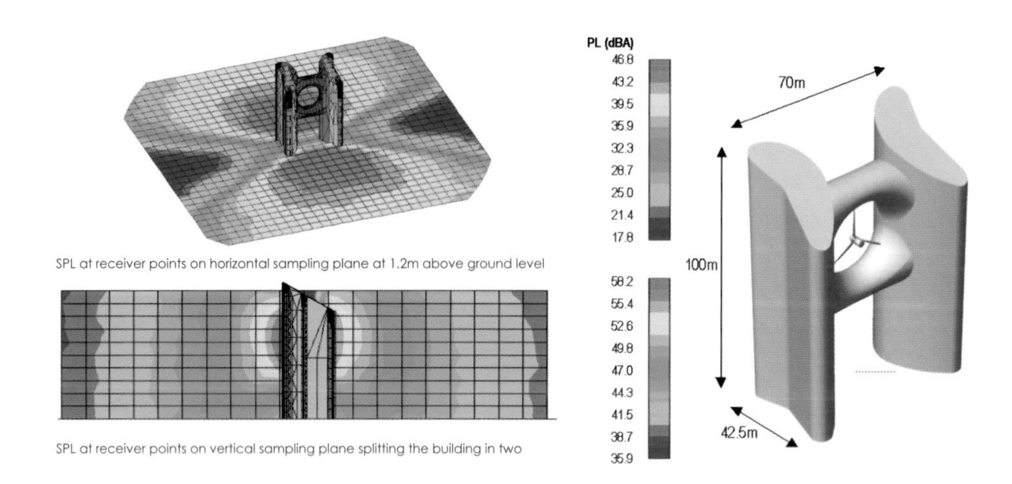

Broad conclusions can be drawn from these computer studies:

Figure 5.14 Example of a sound power level (SPL) propagation simulation (Project WEB – BDSP Partnership)

- A building can simultaneously act in a number of different ways in terms of noise propagation: to concentrate sound waves close to the building; to reflect them away from the building; to shield the surroundings; and to absorb sound energy.
- For integration within an aerodynamic building, the highest SPL levels will occur in directions perpendicular to the plane of rotation of the (HAWT) turbine blades. Infills will shield external receivers directly beneath them, but the SPL levels will typically be slightly higher (4–5 dBA) than for a stand-alone machine along this axis.
- Even for a completely concrete structure, the maximum SPL on a plane 1.2m above ground did not exceed 50dBA for this particular turbine – the strictest noise standard normally applied for the granting of planning permission.

For internal receivers (i.e. occupants within the aerodynamic building), concentration of sound waves will require specialist acoustic treatment of the infills and possibly the office façades facing onto the turbine(s). Absorbent materials such as acoustic plaster, expanded polyurethane foam and fibreboard could be used or additional layers of glazing specified. This would also minimize noise propagation problems, reducing general SPL levels near to the ground by around 3dBA.

Some acoustic measurements were taken during the Project WEB field testing. However, it was difficult to distinguish noise propagated from the small integrated turbines above the background levels and wind noise.

Simulations were also carried out for the prototype building placed in an urban context. The development was assumed in this case to dominate its immediate surroundings. This would be the typical case in order for the turbine to access the greatest wind resources. The minimum (plan) distance to the closest points on these buildings varies between 15 and 25m. The maximum SPL of ~53dBA occurs on the buildings immediately behind the turbine. This is lower than the 55dBA planning noise limit often set for commercial areas.

Specialist acoustic treatment of the façades and infills of associated with a wind turbine will generally be required, certainly for sites in commercial or residential districts.

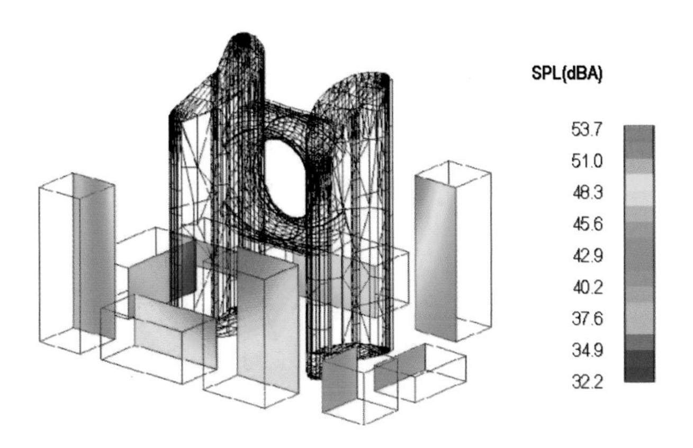

Figure 5.15 Impact of sound power level (SPL) on surrounding buildings (Project WEB – BDSP Partnership)

The acoustic properties of any safety devices incorporated into the building could similarly provide an extremely important absorbing and screening role, since they will be mounted normal to the rotor plane of the turbine. These options may conflict with the architectural desire for transparency, which could only be resolved through detailed design.

A combination of these approaches should make it possible for the noise impact of a building-integrated turbine development to be considered acceptable in planning terms, depending on its cumulative impact in conjunction with background levels.

Architectural integration and organization of space

Aerodynamically optimal designs may prove suboptimal in terms of economic organization of space. Neither will they necessarily produce energy-efficient buildings, particularly if they contain deep floor plates where daylighting and natural ventilation options are restricted.

Concerns over the quality and value of the spaces adjacent to the turbines will inevitably arise due to concerns over noise transmission, flickering of rotating blades and electromagnetic interference with electronic equipment. However, the quality of these spaces can be predicted to a certain extent and improved through the application of various design options such as faced treatments and spatial organization.

A sensible means of spatial organization would be to place intermittently used or service areas (i.e. lifts, stairs, cores) adjacent to the turbine(s), as they have less demanding requirements than normal (office) space and can provide a buffering role. The interiors of infills linking twin-tower buildings can be used as architecturally stimulating transitional spaces or 'sky lobbies'.

Figure 5.16 Conceptual architectural design of a 200m tower with three integrated building turbines with 30m blade diameters (Project WEB - ibk2 University of Stuttgart)

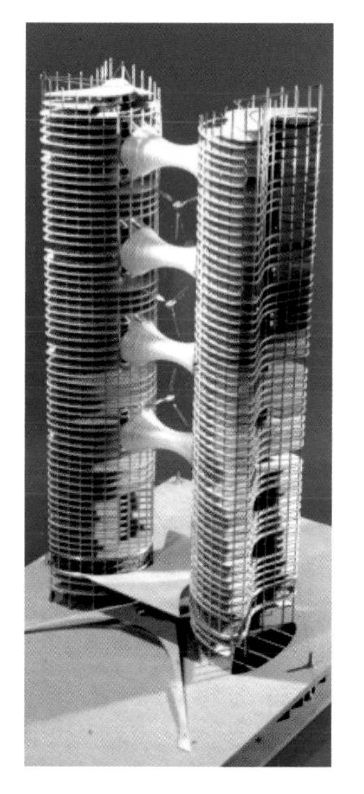

Photograph of architectural model

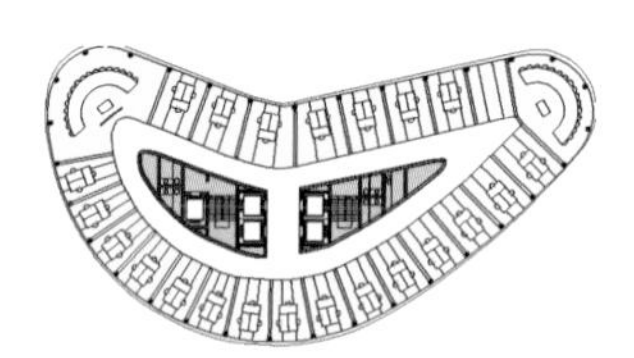

Standard office layout (no special measures)

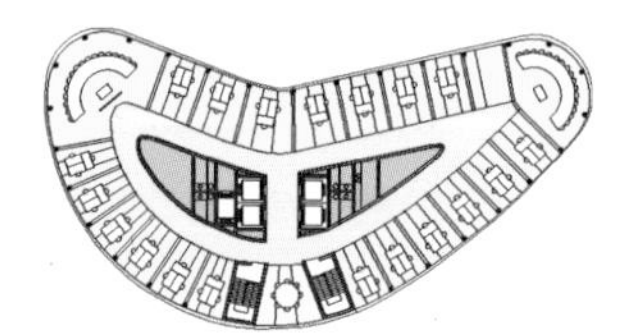

Lifts and lobby act as buffer spaces

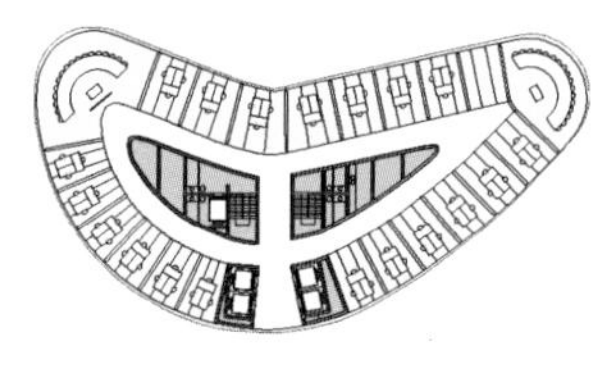

Stairs and small meeting room act as buffer spaces

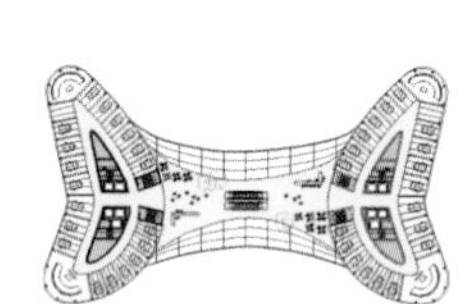

Typical floor plans

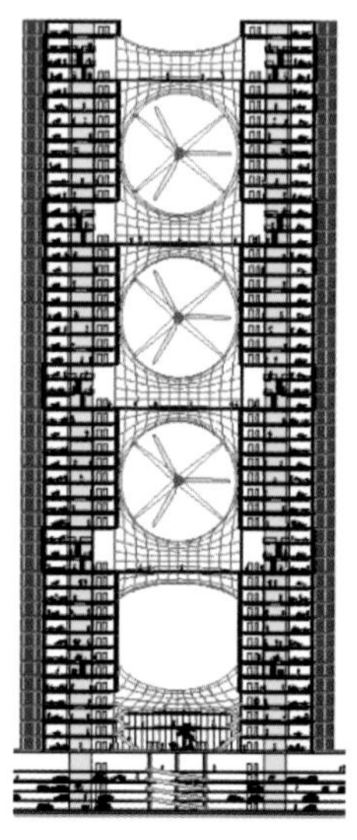

Front elevation

Figure 5.17 The sky lobbies which have views of a 30m blade diameter wind turbine (Project WEB - ibk2 University of Stuttgart)

The prototype 200m twin tower in Figure 5.16 is based on the use of the aerodynamically efficient 'boomerang' footprint. The curved three-dimensional infills link the two symmetrical towers and create an aerodynamic fit around the three integrated HAWTs.

The infill spaces could contain walkways, plants, mezzanine levels, restaurants, bars etc., with services routed underneath the connecting walkways. However, several issues would have to be solved which would conflict with the architectural desire for lightness of structure and transparency:

- For safety and acoustic (vibration) reasons, materials which are good impact energy and sound energy absorbers will be required.
- Noise and aerodynamic considerations (i.e. the need for smooth surfaces on the infills for wind enhancement), will make it difficult to have controllable openings on the infills for supply and extract air.
- Adverse visual impacts for occupants of the building due to the shadow and flickering of the rotating blades might necessitate the installation of opaque façade panels adjacent to the turbines.

The issue of how to best use the electrical power generated by the integrated turbines also has to be addressed. For larger turbines it may be possible to connect directly into the low-voltage supply network in a building plant room. If surplus power is to be exported to the grid, it will be necessary to connect into the local electricity substation supplying the building (normally located outside of the building), where load matching can take place. Providing the distance the cables run is relatively small it should not be necessary to step up and then step down the voltage to avoid transmission losses.

Planning and the effect of local buildings on turbine performance

Building-integrated turbines in particular may be sited very close to other buildings. In ideal circumstances the turbine will be elevated well above the neighbouring buildings. However, there may be cases where a tall building will be adjacent to a turbine and be able to affect the quality of the winds arriving at that turbine.

In most cases building-integrated turbines will be sited so as to minimize the effect from protruding buildings and, as a minimum, ensure that the energy-rich prevailing winds are not diminished.

The extent of any impacts on energy performance will depend on the size of an adjacent buildings and distance it is away from the turbine. However, even if the building appears to be relatively far away, the disruption to the winds flow patterns (e.g. from steady free-stream flow to vortex shedding flow) can last for hundreds of metres. The influence may even extend to reducing the longevity of the turbine if the swirling winds in the wake of the building cause repeated fluctuations and stresses on the turbine, i.e. through wake buffeting.

An interesting and important question that naturally arises from this discussion is the impact of future developments that are proposed next to an existing building-integrated turbine and the extent planners should act to preserve wind resources.

This type of issue also applies to building-integrated solar energy installations and planning conditions already cover issues such as 'rights to lights' and to the prevention of 'overshadowing'. In many cases 'rights to light' and 'overshadowing' computational studies are requested to ensure the impacts of a development on the existing buildings is within acceptable limits. Similarly pedestrian wind comfort studies are often requested during the planning stage to ensure that a proposed development does not deteriorate wind comfort and safety conditions near existing buildings.

It seems that the resolution of this planning issue should therefore be taken on a case-by-case basis and planners in certain instances should request an assessment of the impact on a given turbine from a proposed development.

Smoke tests in wind tunnels can reveal these phenomena and other transient effects such as galloping, flutter and wake interference. Transient CFD simulation can now also predict these phenomena as parallel computing capabilities allow the use of sufficiently refined grids and sufficiently small time steps to capture the driving turbulence phenomena.

SUMMARY

Wind turbines can be integrated into buildings in many forms and can deliver a substantial proportion of the electricity demand if good wind resources are available and the design of the building is energy efficient. Wind energy generation can be improved by using the building form to access higher quality winds and by locally concentrating the winds. The generic forms of integration outlined have been shown to accelerate wind to different degrees depending on the local wind regime. Energy enhancements can produce more than double the amount of energy produced for certain key wind directions and overall energy increases of 10 to 50 per cent (over free-standing equivalent at the same height) for basic, non-optimized forms.

These energy yield benefits can be more than anticipated, due to the fact that appropriately curved façades are able to 'turn' the winds in order to align them with the turbine. The twin tower from Project Web, for example, was able to demonstrate enhanced power output for a wide range of incident wind angles (+/- 75°) onto the building.

Accelerating winds to increase energy yields will produce additional design issues. However, with an awareness of the pitfalls, effective action can counteract any potential negative effects - e.g. ensuring no additional significant heat or 'coolth' losses occur through a façade exposed to faster wind, or mitigating the effect of noise and the visual effect of the blades on the occupant.

Although building-integrated turbines can produce iconic building forms, wind energy integration also extends in multiple building forms. The arrangement of two relatively simple buildings, either side of a turbine, can produce notable energy enhancements. Wind energy integration can also involve more modest and subtle uses, as demonstrated by the horizontal VAWT incorporated into the design of the Mercy Housing Lakefront in Chicago (Figure 5.18).

Figure 5.18 Building-integrated turbines on the Margot and Harold Schiff Residences project by Murphy Jahn Architects using Aerotecture (www.aerotecture.com) horizontal VAWT to feed 96 apartments (with other green features such as solar thermal and rainwater collectors) in North Clybourn Avenue, Chicago (2007)
(Doug Snower Photography, Chicago)

REFERENCES

1 'Council House 2' at www.melbourne.vic.gov.au (July 2008)

2 The Chartered Institution of Building Service Engineers (www.cibseorg)

3 *CIBSE Guide A*, CIBSE, London (1987)

Conclusions

Erecting the Ford Turbines in Dagenham, UK (Ecotricity)

Like any up-and-coming field, urban wind energy has certain requirements if its potential is to be realized. This book goes some way towards addressing several of these needs, such as increasing the confidence levels of key stakeholders and providing useful information to enable timely progression and successful delivery of new projects.

Conclusions

PART 1 placed urban wind energy in the context of key global energy, environmental and economic drivers. Growing environmental awareness and an increasing demand for solutions capable of helping bring about the transformation from dependency on fossil fuels to a sustainable energy future bodes well for the success of this sector, especially given predicted energy price rises and that large-scale turbines become 'carbon positive' within 3 to 12 months of their operation.

PART 2 explored the potential of wind energy to fit into the urban environment in order to go part of the way to meet the need for clean, secure renewable energy. A wide range of options are possible and, despite cautious beginnings, there are now a growing number of successful, innovative projects being carried out. Some of these involve large-scale turbines, with blade diameters of 80m or more, operating very close to buildings (one blade diameter away), although the distance from residential buildings has to be carefully considered.

PART 3 focused on how to assess the feasibility of a particular project in energy, environmental and economic terms to ensure that good design decisions are made. It should be clear that wind energy, unlike other renewable technologies such as biomass boilers, and even PV to some extent, is highly dependent on specific local factors (including the planning system and public opinion). Therefore, what may be appropriate or feasible at one location may be completely unsuitable somewhere else even if superficially similar. In many cases, associated environmental impacts can be satisfactorily mitigated if due consideration has been given to the various aspects described in this text. The need for on-site wind monitoring (resource prospecting) has also been reiterated.

PART 4 emphasized practical design aspects and presented an overview of currently available technology. This technology has evolved via a certain route producing only a limited number of reliable 'off the shelf' options. The quality of manufacturing for many large-scale turbines is now high and the availability is typically greater than 97 per cent. However, with some smaller turbine designs further improvements will be required to improve product reliability, deal with turbulence, reduce costs, and improve longevity – to realize their full potential in urban environments.

Finally, **PART 5** examined building-integrated wind turbines (where turbine integration drives the form of the building) and identified some of the key design aspects involved in these more ambitious proposals. As sustainability moves up the list of priorities of more and more design briefs, development teams will look to move renewable energy integration into the core of their design practices. Meanwhile, the necessary skills, experience and products from manufacturers are also developing. Architects and developers, tuned in to the need for sustainable design or responding to the growing awareness of the public, can now regard integrated wind energy as a worthwhile option to consider as part of a holistic design response.

If urban wind energy is to develop further there are several key groups of actors who can make significant contributions:

- **Developers, investors, businesses, home-owners.** Their interest and commitment – not just to wind energy but also to sustainable development, energy efficiency and renewable energy – can set examples for others to follow.

- **Architects and engineers.** Their action in proposing this technology where appropriate and presenting a balanced, reasoned case as an integral part of the overall design concept is vital. Innovative, Imaginative designs can create architecture and urban landscapes that can inspire and bring about change.

- **Manufacturers and industry groups.** Their investment in the future to provide best possible technology (while reducing costs) will generate useful mechanisms to support customers interested in integrating wind energy into the built environment. Going forward, setting standards for testing, performance reporting and safety for small wind turbines (<50kW) can help to create a broader international market.

- **Planners and policy-makers.** Their investment in resources can speed up the assessment of proposals and improve guidance. This may include defining new roles in departments (such as a Renewable Energy Facilitator) and recognizing urban wind energy in energy policies and funding processes. At national and international level, wider recognition as a distinct technology area in energy policy and planning (including in financial incentives for encouraging development of renewable energy) will be important.

- **Students.** Their fresh view may help with the transition from traditional architectural/ engineering design approaches to more sustainable forms.

Wind energy has a key role to play in the future as part of a diverse portfolio of renewable energy technologies and energy-efficient practices. In particular, if energy storage issues can be addressed in the future (e.g. using hydrogen, via pumping and damming of water, or through electrochemical means), a significant proportion of the world's energy needs can be met in a clean and safe manner. Wind turbines in prominent urban locations can not only generate energy but also help individuals keep in mind the balance between man and nature, as well as the importance of combining renewable energy generation and energy efficiency in moving towards sustainable societies. Moreover, if this technology can be integrated into our buildings and the urban landscape in a meaningful and appropriate manner, it may capture our imaginations and begin to change mindsets. As our energy future is certain to rely increasingly on a multitude of renewable sources, wind energy has a place in the built environment for some time to come.

LARGE-SCALE TURBINES

- Vestas (Denmark) www.vestas.com
- GE (US) www.gepower.com
- Enercon (Germany) www.enercon.de
- Gamesa Eolica (Spain) www.gamesa.es/en
- Neg Micon (now part of Vestas)
- REpower (Germany) www.repower.de
- Nordex (Germany) www.nordex-online.com
- Suzlon (India) www.suzlon.com/WindTurbines.html
- Acciona (Spain) www.acciona.es
- Ecotecnia (Spain) www.ecotecnia.com
- Siemens Wind Power (formerly Bonus) (Germany) www.bonus.dk, www.powergeneration.siemens.com
- MHI Mitsubishi Heavy Industries (Japan) www.mhi.co.jp
- DeWind (Germany) www.compositetechcorp.com/windpower.htm
- Goldwind (China) cn.goldwind.cn
- Scanwind (Sweden) www.scanwind.com
- Clipper Windpower (US), www.clipperwind.com
- Emergya Wind Technologies (formerly Lagerwey) (The Netherlands) www.directwind.nl

MID-RANGE TURBINES

- Distributed Energy Systems, Northern Power (US) www.distributed-energy.com or www.northernpower.com (NPS 100 21m/100kW)
- Fuhrländer AG (Germany) www.fuhrlaender.de (FL 25kW, FL 30kW, FL 100kW)
- Subaru or Fuji Heavy Industries FHI (Japan) www.fhi.co.jp
- ACSA Aerogeneradores Canarios S.A (Spain) www.acsaeolica.com (A27/225kW)
- Turbowinds (Belgium) www.turbowinds.com (T400-34, T600-48)
- Norwin (Denmark) www.normin.dk (29m/225 kW, 47m/750 kW)
- NEPC (India) www.nepcindia.com (40kW, 50kW, 100kW, 180kW)
- The Entegrity Wind Systems Inc. (formerly Atlantic Orient Corporation) www.entegritywind.com (EW50, formerly AOC 15/50)
- Energie PGE (Canada) www.energiepge.com/default.php?langue=en (20m/50kW)
- Bergey Windpower Co (US) www.bergey.com (BWC EXCEL - 10 kW),
- Gaia-Wind Ltd (US) www.Gaia-Wind.com (13m/11kW)
- WES Wind Energy Solutions (Canada) www.windenergysolutions.ca WES 18 (18m/80 kW & WES 30 (30m/250 kW)
- Wind Turbine Industries Corp (US) www.windturbine.net 23-10 Jacobs (10 kW), 31-20 Jacobs (20 kW)
- Westwind (UK) www.westwindturbines.co.uk (10kW, 20kW)
- Pitchwind (Sweden) www.pitchwind.se (20kW)
- Proven (UK) www.provenenergy.co.uk (15kW)

MICRO/SMALL TURBINES (<10KW)

- Proven (UK) www.provenenergy.co.uk (0.6kW, 2.5kW, 6kW)
- Westwind (UK) www.westwindturbines.co.uk (3kW, 5kW)
- Eoltec (France) www.eoltec.com (Scirocco 5.6m/6kW)
- Iskra (UK) www.iskrawind.com (5.4m/5.3kW)
- Eclectic energy (UK) www.eclectic-energy.com (1.1m/0.4kW)
- ACSA Aerogeneradores Canarios S.A (Spain) www.acsaeolica.com
- Bergey Windpower Co. (US) www.bergey.com (BWC XL 1kW)
- Wind Energy Solutions (The Netherlands) www.windenergysolutions.nl (Tulipo, 5m /2.5kW)
- Southwest Windpower Co. (US) www.windenergy.com AIRX (400W), Whisper (900W, 1 kW, 3 kW), Skystream 3.7m (1.8 KW)
- Quantum Wind (Canada) www.quantumwind.com (5m/5kW)
- Windsave (UK) www.windsave.com (1.9m/1.25kW)
- Renewable Devices (UK) www.renewabledevices.com (Swift 2m/1.5kW)
- Marlec (UK) www.marlec.co.uk (250W)

SMALL VAWT (<10KW)

- Quietrevolution (UK) www.quietrevolution.co.uk (qr5, 6kW)
- Winside (Finland) www.windside.com (up to 5kW)
- Ropatec (Italy) www.ropatec.com (300W, 1kW, 3kW, 6kW, 20kW)
- Cleanfield Energy (Canada) www.cleanfieldenergy.com (3.5kW)
- Turby (The Netherlands) www.turby.nl (2.5kW)
- Vertical Wind Energy (UK) www.vweltd.com (3.6kW)
- Helix Wind (US) www.helixwind.com (5kW)
- Windterra (US) www.windterra.com (1.2kW)
- Mariah Power (US) www.mariahpower.com (1.2kW)

Great Lakes Science Centre Turbine, Cleveland, USA (Jim Kolmus)

Acronyms and abbreviations

BWEA	British Wind Energy Association
CFD	computational fluid dynamics
CH_4	methane
CHP	combined heat and power
CO_2	carbon dioxide
Cp	coefficient of performance
CSS	carbon capture and storage
DTI	Department of Trade and Industry
EIA	environmental impact assessment
HAWT	horizontal axis wind turbine
IPCC	Intergovernmental Panel on Climate Change
NOABL	Numerical Objective Analysis of Boundary Layer
NO_x	nitrogen oxides
NPV	net present value
PV	photovoltaic
ROC	Renewable Obligation Certificate
ROI	return on investment
SO_x	sulphur oxides
SPL	sound pressure level
SST	shear stress transport model
SWL	sound power level
TSR	tip speed ratio
VAWT	vertical axis wind turbine
VRB	vanadium redox battery
WEB	Wind Energy in the Built Environment Project
ZED	Towards Zero Emission Urban Development Project